MW00355068

The Origin of Universes: of Quaternion Unified SuperStandard Theory (QUeST); and of the Octonion Megaverse (UTMOST)

Stephen Blaha Ph. D.
Blaha Research

Derivation of QUeST from a One Dimension – One *Megaverse* Fermion Theory
Origin of Universes
Derivation of UTMOST from a One Dimension – One 10 Dimension Fermion Theory
Origin of the Megaverse
Patterns of Symmetry Breaking of QUeST
Derivation of UST from QUeST
Interconnection of Our Universe and the Megaverse
Overview of QUeST and its UST Sector Features
Overview of Megaverse UTMOST Features

Pingree-Hill Publishing
MMXX

Copyright © 2020 by Stephen Blaha. All Rights Reserved.

This document is protected under copyright laws and international copyright conventions. No part of this book may be reproduced, stored in a retrieval system, or transmitted by any means in any form, electronic, mechanical, photocopying, recording, or as a rewritten passage(s), or otherwise, without the express prior written permission of Blaha Research. For additional information send an email to the author at sblaha777@yahoo.com or call 603-289-5435.

ISBN: 978-1-7356795-0-1

This document is provided "as is" without a warranty of any kind, either implied or expressed, including, but not limited to, implied warranties of fitness for a particular purpose, merchantability, or non-infringement. This document may contain typographic errors, technical inaccuracies, and may not describe recent developments. This book is printed on acid free paper.

Rev. 00/00/01 September 21, 2020

To Margaret

Some Other Books by Stephen Blaha

All the Megaverse! Starships Exploring the Endless Universes of the Cosmos using the Baryonic Force (Blaha Research, Auburn, NH, 2014)

SuperCivilizations: Civilizations as Superorganisms (McMann-Fisher Publishing, Auburn, NH, 2010)

All the Universe! Faster Than Light Tachyon Quark Starships & Particle Accelerators with the LHC as a Prototype Starship Drive Scientific Edition (Pingree-Hill Publishing, Auburn, NH, 2011).

Unification of God Theory and Unified SuperStandard Model THIRD EDITION (Pingree Hill Publishing, Auburn, NH, 2018).

The Exact QED Calculation of the Fine Structure Constant Implies ALL 4D Universes have the Same Physics/Life Prospects (Pingree Hill Publishing, Auburn, NH, 2019).

Unified SuperStandard Theory and the SuperUniverse Model: The Foundation of Science (Pingree Hill Publishing, Auburn, NH, 2018).

Quaternion Unified SuperStandard Theory (The QUeST) and Megaverse Octonion SuperStandard Theory (MOST) (Pingree Hill Publishing, Auburn, NH, 2020).

Unified SuperStandard Theories for Quaternion Universes & The Octonion Megaverse (Pingree Hill Publishing, Auburn, NH, 2020).

The Essence of Eternity: Quaternion & Octonion SuperStandard Theories (Pingree Hill Publishing, Auburn, NH, 2020).

A Very Conscious Universe (Pingree Hill Publishing, Auburn, NH, 2020).

Hypercomplex Universe (Pingree Hill Publishing, Auburn, NH, 2020).

Why is the Universe Real? From Quaternion & Octonion to Real Coordinates

Available on Amazon.com, bn.com Amazon.co.uk and other international web sites as well as at better bookstores (through Ingram Distributors).

CONTENTS

INTRODUCTION 1

1. BIRTH OF THE QUEST UNIVERSE 3

1.1 BQUEST ORIGIN 3
1.2 DYNAMICS OF THE SEED FERMION 3
1.3 INDICES AND DIMENSIONS 4
1.3.1 The Difference between Indices and Dimensions 4
1.4 ORIGIN OF UNIVERSE 5

APPENDIX 1-A. RECAP OF FEATURES OF QUEST QUATERNION SPACE 7

1-A.1 THIRTY-TWO COMPLEX QUATERNION SPACE – 32 × 8 DIMENSION ARRAY 7
1-A.1.1 QUeST Internal Symmetry Groups 8
1.4.1.2 QUeST Space-Time 8
1-A.1.3 Generation and Layer groups of UST, QUeST and BQUeST 8
1-A.1.4 Fermion- Dimension Duality 8
1-A.1.5 Fermion Structure Extracted from QUeST Symmetry Structure 8
1-A.2 THIRTY-TWO COMPLEX QUATERNION SPACE – 16 × 16 DIMENSION ARRAY 13
1-A.3 A PARTITION TO REAL 3+1 DIMENSION SPACE-TIMES 19

APPENDIX 1-B. EVIDENCE FOR UNIVERSE PARTICLES 21

1-B.1 EVIDENCE FOR ENTITIES BEYOND THE UNIVERSE 21
1-B.1.1 Great Attractors 21
1-B.1.2 Bright Bumps in Universe Suggesting Collision with Another Universe 21
1-B.1.3 Cold Spot in Universe Suggesting Collision with Another Universe 22
1-B.1.4 Megaverse Energy-Matter Infusion into Our Universe 22
1-B.1.5 Conclusion 22
1-B.2 HUBBLE CONSTANT AND UNIVERSE EXPANSION 22
1-B.2.1 Hubble Constant Experimental Data 22
1-B.2.2 Fit to the Hubble Constant Data and Scale Factor 23
1-B.3 HUBBLE PARAMETER AND VACUUM POLARIZATION OF A PARTICLE 24
1-B.3.1 Vacuum Polarization Generation of the Early Time Part of the Universal Scale Factor 24
1-B.3.2 Comparison of QED Vacuum Polarization Exponent with Universe Vacuum Polarization Exponent 25
1-B.3.3 A New Vector Interaction for Universe Particles 25
1-B.3.4 Second Order Vacuum Polarization of a Scalar Universe Particle 26
1-B.3.5 Finding the Universe g_U 27
1-B.3.6 Dark Energy is Equivalent to Universe Vacuum Polarization 29
1-B.3.7 Quasi-Free Universe Particles 29
1-B.3.8 Doubling Relation Between Coupling Constants 29

APPENDIX 1-C. PSEUDOQUANTUM FIELD THEORY 31

2. UST DERIVED FROM QUEST 33

2.1 INTERNAL SYMMETRIES 33
2.2 QUEST → UST SYMMETRIES 36
2.3 U(4) GENERATION, U(4) LAYER AND U(1) FERMION GROUPS 38
2.4 THE GENERATION GROUP 38

2.5 The Layer Group 38
2.6 The Fermion Groups 39
2.7 Particle Spectrums of QUeST: Fermions, and Vector Bosons 40
2.7.1 Vector Bosons *40*
2.7.2 Fermions *40*

3. FUNDAMENTAL QUEST-UST AXIOMS 43

3.1 Fundamental Prerequisites for a Fundamental Theory of Physics 43
3.1.1 Particles, Quantum Field Theory, Quaternions, Higgs Particles *44*
3.2 UST Axioms 45
3.2.1 The Derivation of the Unified SuperStandard Theory *46*
3.3 New Deeper Axioms for QUeST-UST 47
3.4 General Implications of the QUeST-UST Axioms 48

4. ENHANCED QUANTUM FIELD THEORY FROM CHAPTERS 40 AND 41 OF BLAHA (2020C) 51

40.1 Two-Tier Features in 4-Dimensional Space-Time 51
40.2 Simple Two-Tier X^{μ} Formalism 53
40.3 Y^{μ} Gauge 53
40.4 Scalar Field Quantization Using X^{μ} 56
40.5 Scalar Feynman Propagators 57
40.6 String-like Substructure of the Theory 58
40.7 Two-Tier Complexon Quantum Fields 59
40.8 Complexon Feynman Propagator 61
40.9 Vacuum Fluctuations 61
40.10 Time Intervals in General Relativity 62
40.11 Vacuum Fluctuations in the Gravitation Fields 63
40.12 Two-Tier Features in D-Dimensional Space-Time (such as the Megaverse) 63
41.1 General Case of PseudoQuantization in Differing Coordinate Systems 67
41.2 Two-Tier PseudoQuantum Field Theory 69
41.3 PseudoQuantum Higgs Scalar Particle Field Theory in D-dimensional Space-Time 70
41.3.1 The Enigma of Higgs Particles and the Higgs Mechanism *70*
41.3.2 PseudoQuantization of Scalar Particles *72*
41.3.3 Vacuum States for Scalar (Higgs) Particles with Non-Zero Vacuum Expectation Values *73*
41.3.4 Interpretation of Negative Energy Scalar Particle States *74*
41.3.5 Contrast with Conventional Second Quantization of Scalar Particles *75*
41.3.6 Why Inertial Reference Frames are Special *76*
41.3.7 PseudoQuantization Reveals More Physical Consequences than the Higgs Mechanism of Scalar Particles *76*
41.3.8 The T Invariance Issues of Our PseudoQuantized Scalar Particle Theory *77*
41.3.9 Retarded Propagators for Our Quantized Higgs Particles *77*
41.3.10 The Local Arrow of Time *78*
41.3.11 Space-Time Dependent Particle Masses *79*
41.3.12 Inertial Mass Equals Gravitational Mass *80*
41.3.13 Benefits of the PseudoQuantization Method *81*
41.1 Two-Tier Formulation and PseudoQuantization in the Megaverse 81

5. QUEST-UST LAGRANGIAN FORMULATION BASED ON THE RIEMANN-CHRISTOFFEL CURVATURE TENSOR 83

5.1 The Covariant Derivative 83
5.2 The Curvature Tensor 84
5.3 Vector Boson and Graviton Lagrangian Terms 90

5.4 NEW VECTOR BOSON INTERACTIONS 92
6. OTHER QUEST-UST TOPICS 95
7. ORIGIN OF MEGAVERSE 97
7.1 BMOST ORIGIN 97
7.2 DYNAMICS OF THE URFERMION 97
7.3 INDICES AND DIMENSIONS 98
7.3.1 The Difference between Indices and Dimensions *98*
7.4 ORIGIN OF MEGAVERSE 99
APPENDIX 7-A. FEATURES OF UTMOST OCTONION SPACE 101
7-A.1 UTMOST SPACE 101
7-A.2 UTMOST FERMIONS 106
7-A.3 PARTITION OF UTMOST INTO QUEST SUBSPACES 108
8. THE CONNECTION OF QUEST AND UTMOST 111
REFERENCES 113
INDEX 121
ABOUT THE AUTHOR 125

FIGURES and TABLES

Figure 1.1. Diagram for the transition of the seed fermion to a universe particle........... 5
Figure 1-A.1. The 32 complex quaternion dimension QUeST array. This array is the 32 × 8 array of •'s. 7
Table 1.1. Map between fundamental representations and their dimensions................. 8
Figure 1-A.2. The 32 complex quaternion dimension QUeST array subdivided into 4 layers of 8 rows..Each layer will be seen to map to a block of fundamental group representations as shown in Figs. 1-A.3 and 1-A.4.. 9
Figure 1-A.3. The four layers of QUeST internal symmetry groups (and space-time) for 32 dimension complex quaternion space. Note: each row has an 8 • complex quaternion. Note the left column of blocks combine to specify a 4 dimension complex quaternion space-time. Note each layer requires 64 dimensions... 10
Figure 1-A.4. The internal symmetry groups for one QUeST layer of the 4 layers in the 32 × 8 dimension array format. The two large blocks are each 5 complex dimension (10 real dimension) representations of SU(2)⊗U(1)⊗SU(3). The U(2) group (badly broken) supports transformations (rotations) between Normal and Dark matter. 11
Figure 1-A.5. Fermion particle spectrum and partial examples of the pattern of mass mixing of the Generation group and of the Layer group. Unshaded parts are the known fermions with an additional, as yet not found, 4th generation. The lines on the left side (only shown for one layer) display the Generation mixing within each layer. The Generation mixing occurs within each layer using a separate Generation group for each layer. The lines on the right side show Layer group mixing (for Dark matter) with the mixing among all four layers for each of the four generations individually. There are four Layer groups for Normal matter and four Layer groups for Dark matter.. There are 256 fundamental fermions. QUeST and UST have the same fermion spectrum........... 12
Figure 1-A.6. Fundamental fermions have a 1:1 correspondence with QUeST dimensions. Note the number of dimensions in each row is 8 – the number of dimensions in a complex quaternion. Correspondingly the number of fermions in each row is 8 – a suggestive similarity. Each layer has four normal fermion generations and four Dark fermion generations. Each dot (pebble) represents a dimension in the left part of the figure and a fermion in the right part.. 13
Figure 1-A.7. The 16 × 16 array of QUeST dimensions. 14
Figure 1-A.8 The 16 × 16 form of QUeST array has two "layers", each of which is composed of two layers of the 32 × 16 four layer figure of Fig. 1-A.3. Note each of these "layers" has 128 dimensions. 15
Figure 1-A.9. Set of 16 dimension blocks for the first "layer" (of two "layers") of the 16 × 16 array. Each "dashed" block (regardless of its apparent size) is a 4 × 4 = 16 array of

dimensions. This set of 8 blocks contains the 8×16 = 128 dimensions of "layer" 1. "Layer" 2 is similar. The Dark U(2) groups supports transformations (rotations) between the types of matter: Normal and Dark1; and Dark2 and Dark3. 16
Figure 1-A.10. The two "layers" of 4 × 4 dimension subblocks of the 16 × 16 dimension array. 17
Figure 1-A.11. Spectrum of the generations of fermions of QUeST for the 16 × 16 dimension array representation. Each fermion is represented by a •. Quark triplets are represented by three •'s. Note there are 256 fundamental fermions............................ 18
Figure 1-A.12. Block form of a 16 × 16 QUeST fermion array with each block row corresponding to one layer. Each block contains four generations of fermions. The result is 4 × 4 blocks. The label e q-up indicates a charged lepton – up-type quark pair, ν q-down indicates a neutral lepton – down-type quark pair, and so on. Note the blocks can be reaaranged into a 32 × 8 form without physical consequences at this level of discussion since the right two columns and the lowest two rows are all Dark at present. 18
Figure 1-A.13. The 32 complex quaternion dimension QUeST array partitioned down to a real spwce-time. The partition labeled "1" reduces the array to 32 quaternion dimensions after discarding the right columns. The partition labeled "2" reduces the array to 32 complex dimensions similarly. The partition labeled "3" reduces the array to 32 real-valued dimensions in the leftmost column.. 19
Figure 1-A.14. Spectrum of the generation of fermions. Note only one layer of fermions. Two leptons and 6 quarks in each row of Normal matter. Similarly for Dark matter. 20
Figure 1-B.1 One loop vacuum polarization boson Feynman diagram. 26
Figure 1-B..2 The interaction coupling constants show a regular doubling. A fundamental cause for doubling is not apparent. 30
Figure 2.1. Pattern of symmetry breaking of QUeST U(128). 33
Figure 2.2. The four layers of QUeST internal symmetry groups (and space-time) for 32 dimension complex quaternion space. This is a changed Fig. 1-A.3 (U(2) → *U(1)⊗U(1))*. It is done to support the symmetry breakdown pattern above. Note: each row has an 8 • complex quaternion. Note the left column of blocks combine to specify a 4 dimension complex quaternion space-time. Note each layer requires 64 dimensions. 35
Figure 2.3. Four layers of Internal Symmetry groups in QUeST. The groups in each layer are independent of those in other layers. The groups in each block of each layer are independent of those in the other blocks. Each block contains 16 dimensions. The dimensions furnish fundamental representations for the groups listed. The entire set of blocks contains 256 dimensions. Each layer contains 64 dimensions. The first two columns are for the "Normal" sector. The last two columns are for the "Dark" sector

(although most of the Normal sector is Dark observationally at present.) This Figure also holds for UST with the addition of Fermion groups. 36
Figure 2.4. Four layers of Internal Symmetry groups in UST from Blaha (2020c) and earlier books such as Blaha (2018e). The groups in each layer are independent of those in other layers. The groups in each block of each layer are independent of those in the other blocks. The first two columns are for the "Normal" sector. The last two columns are for the "Dark" sector (although most of the Normal sector is Dark observationally at present.) Note columns 1 and 3 of UST omit a U(1) group relative to Fig. 2.3. Also space-time is separate and not included with of the set of symmetries in UST. The UST tiles in columns 1 and 3 have 10 dimensions; the tiles in columns 2 and 4 have 16 dimensions. 37
Figure 40.1. Feynman diagram for conventional and the nth diagram of a cloaked Two-Tier propagator. 59
Figure 7.1. Diagram for the transition of the urfermion to a Megaverse........................ 99
Figure 7-A.1. The 64 complex octonion dimension UTMOST array. This is the 64 × 16 array of •'s. It has 1024 dimensions. 101
Figure 7-A.2. The UTMOST array with 32 × 32 dimensions for a 32 quaternion octonion dimension space.. 102
Table 7-A.1. Map between fundamental representations and their dimensions........... 102
Figure 7-A.3. The internal symmetry groups of *one layer* (consisting of 8 rows in Fig. 7-A.2) of the four layers of 32 × 32 dimension UTMOST. The other three layers are copies of the this layer. Note the Dark U(4) groups. One U(4) "rotates" among Normal, Dark1, Dark2, and Dark3. The other U(4) "rotates" among Dark4, Dark5, Dark6, and Dark7. 103
Figure 7-A.4 The *first* of the four layers of UTMOST dimensions with boxes around sets of dimensions for fundamental group representations. The U(4) Dark groups have been separated into U(2)⊗U(2) factors for later use. 104
Figure 7-A.5. Four layers (each in two rows) in the 32 × 32 dimension UTMOST array composed of 4 × 4 blocks, which are within the four block 8 × 8 sections for each pair: Normal+Dark1, Dark2+Dark3, Dark4+Dark5 and Dark6+Dark7. In total they form the 32 × 32 = 1024 UTMOST dimension array.. 105
Figure 7-A.6. Spectrum of UTMOST fermions in a 16×64 format. Each fermion is represented by a •..Each set of eight •.'s represents a charged lepton, a neutral lepton, three up-type quarks, and three down-type quarks. There are eight sets of four species in four generations which are in turn in 4 layers. There are 1024 fundamental fermions taking account of quark triplets. 106
Figure 7-A.7. Block form of the 32 × 32 UTMOST fermion array with each row corresponding to *half of an UTMOST layer*. Thus 8 × ½ = 4 layers results. Each block contains four generations of fermions. The result is sixty-four 4 × 4 blocks. The label e q-up indicates a charged lepton – up-type quark pair, ν q-down indicates a neutral

lepton – down-type quark pair, and so on. *The form displayed here may explain why generations come in fours.* 107
Figure 7-A.8. The partition of the 32 × 32 dimension UTMOST array into MOST subspaces The size of each subspace is 32×16 = 512 dimensions. 108
Figure 7-A.9. Partition of spectrum of UTMOST fermions in a 16×64 format. Each fermion is represented by a •. Including each quark. Each set of eight •.'s represents a charged lepton, a neutral lepton, three up-type quarks, and three down-type quarks. There are eight sets of four species in four generations which are in turn in 4 layers. There are 512 fundamental fermions in each subspace taking account of quark triplets. Note: Quark singlets won't do; triplets are required. 109
Figure 7-A.10. Partition of block form of the 32 × 32 UTMOST fermion array with each row corresponding to *half of an UTMOST layer*. Thus 8 × ½ = 4 layers results. Each block contains four generations of fermions. The result is sixty-four 4 × 4 blocks. The label e q-up indicates a charged lepton – up-type quark pair, ν q-down indicates a neutral lepton – down-type quark pair, and so on. 110

INTRODUCTION

The Quaternion Unified SuperStandard Theory (QUeST) is a theory for universes that is based on a thirty-two complex quaternion space. It implies the Unified SuperStandard Theory (UST) of the author to the author's initial surprise. The Megaverse theory UTMOST (Megaverse Octonion SuperStandard Theory) is based on a sixty-four complex octonion space. Both theories can be based on deeper one dimension theories.

This book provides a detailed derivation of QUeST from a One Dimension – One Megaverse fermion theory and describes the origin of universes as particles. It also provides a detailed derivation of UTMOST from a One Dimension – One 10 Dimension fermion theory and provides details of the origin of the Megaverse as a particle. One can see that we have reduced the origin of universes and of the Megaverse to the simplest basis. An important aspect of these derivations is the interrelation of universes with the Megaverse. The one fermion seed for universes is in the Megaverse, and the one fermion seed for the Megaverse is in a ten dimension space that may be a SuperString space.

This book also proposes a clear multilevel sequential patterns of symmetry breaking of QUeST down to the level of SU(3) and SU(2)⊗U(1) symmetries. It also considers the detailed derivation of UST from QUeST, and provides an .Overview of QUeST and its UST sector features. An Overview of Megaverse UTMOST features is also presented.

This book completes the panorama of hypercomplex features for QUeST universes and the UTMOST Megaverse. The result is a complete theoretical framework for Elementary Particles and Gravitation.

1. Birth of the QUeST Universe

This chapter describes the process for the birth of a universe from a one dimension – one fermion origin. It describes the BQUeST theory developed in earlier books by the author in 2020. *The 256 dimension QUeST universe arises from one fermion residing in an 8-dimension space, the Megaverse. The fermion has one internal dimension that grows to the 256 internal dimensions of the QUeST universe.* Appendix 1-A describes features of the QUeST quaternion space.

1.1 BQUeST Origin

BQUeST assumes a one dimension space and one fermion. The fermion is introduced to implement the fermion-dimension duality found in QUeST. The fermion has a one *internal* dimension space.

The derivation of the $16 \times 16 = 256$ dimension QUeST array requires the fermion, which we call the *seed fermion*, to be an 8 dimension fermion. The 16 spinor components of the 8 dimension fermion will be used to generate the 16×16 dimension array.

The 8 dimension nature of the fermion raises the question of the space, within which it resides. In Blaha (2020i) and earlier books we showed that the UTMOST theory of the Megaverse has an 8 dimension space-time. We therefore assume that the fermion resides in the Megaverse. We obtain the picture of a universe particle—the fermion—that acts as the seed of the subsequent universe. The evolution (perhaps instantaneous) of the universe begins with a seed of great energy, becomes the 256 dimension QUeST universe in a Big Bang state, then generates internal symmetries and the space-time of QUeST with symmetry breakdown, and then expands from the Big Bang state. Appendix 1-B describes the close analogy between the expansion of the universe and the vacuum polarization of a particle. It strongly suggests that the universe evolved from a type of particle to its present condition.

1.2 Dynamics of the Seed Fermion

The seed fermion must have a dynamics that enables it to generate the 256 dimension array. Since the target array is not symmetric the PseudoQuantum formulation[1] of Quantum Field Theory will be seen to be required.

We therefore define the electromagnetic-like model lagrangian with the seed fermion represented by *two* quantum fields[2] ψ_1 and ψ_2:

$$\mathscr{L} = F^{1\mu\nu} F^{2}{}_{\mu\nu} + \overline{\psi}_2\gamma^0(i\gamma\cdot\partial - m)\psi_1 + \overline{\psi}_1\gamma^0(i\gamma\cdot\partial - m)\psi_2 - e_0\overline{\psi}_2\gamma^0\gamma\cdot A_2\psi_1 - \overline{\psi}_1\gamma^0\gamma\cdot A_1\psi_2 \tag{1.1}$$

where m is the mass-energy of the produced universe, and

---

[1] S. Blaha, Il Nuovo Cimento **49A**, 35 (1979). Reproduced as Appendix 1-B below.

[2] We use a PseudoQuantum Electromagnetic-like pair of fields also. The electromagnetic-like particle is a model of the universe. See eqs. 61 – 90 in Appendix 1-C.

$$F^{i}_{\mu\nu} = \partial_{\nu}A_{i\mu} - \partial_{\mu}A_{i\nu} \tag{1.2}$$

The universe particle is represented by two fields: A_1^{μ} and A_2^{μ}.
The dynamical equations in the Lorentz gauge are

$$\Box A_{i\mu} = e_0 J_{i\mu} \tag{1.3}$$

where

$$J_{1\mu} = \overline{\psi}_2\gamma^0\gamma_\mu\psi_1 \tag{1.4}$$

$$J_{2\mu} = \overline{\psi}_1\gamma^0\gamma_\mu\psi_2 \tag{1.5}$$

The currents have the conservation laws:

$$\partial^{\mu}J_{i\mu} = 0$$

1.3 Indices and Dimensions

Eqs. 1.2 – 1.5 hold for the 8 dimension seed fermion in the Megaverse. If we take the seed fermion off-shell then we can separate eqs. 1.3 – 1.5 into their respective spinor components:

$$\Box A_{i\mu}{}^{ab} = e_0 J_{i\mu}{}^{ab} \tag{1.6}$$

where

$$J_{1\mu}{}^{ab} = \overline{\psi}_1{}^{a}\gamma^0\gamma_\mu\psi_2{}^{b} \tag{1.7}$$

$$J_{2\mu}{}^{ab} = \overline{\psi}_2{}^{a}\gamma^0\gamma_\mu\psi_1{}^{b} \tag{1.8}$$

The (conserved) charge densities ($\mu = 0$) are

$$J_{10}{}^{ab} = \overline{\psi}_1{}^{a}\psi_2{}^{b} \tag{1.9}$$

$$J_{20}{}^{ab} = \overline{\psi}_2{}^{a}\psi_1{}^{b} \tag{1.10}$$

The independence of the spinor components of each of the two seed fermion fields guarantees an array of independent indices, and thence an array of independent dimensions. .

1.3.1 The Difference between Indices and Dimensions

Clearly $A_{i0}{}^{ab}$ is a 16×16 array since the seed fermion exists in an 8 dimension space (the Megaverse) and has 16 component spinors. The array is not symmetric because we use a PseudoQuantum framework.

The array is an array of indices. We now consider the relation of indices and internal dimensions of an entity. First we consider the case of a cube, which can be viewed as an entity with three indices. Its external indices can be taken to label the length, width and height of the cube. Within the cube we can establish a coordinate system with three dimensions labeling points within it. Thus external indices can be viewed as mapping to dimensions within.

Perhaps a better example is a Black Hole. It exists in 4-dimension space. Within its event horizon the coordinates of the external space can be continued. It is also possible to replace them with an internal coordinate system of four dimensions. This possibility could be supported by the role of radial dimension externally becoming effectively a time coordinate internally.

We thus conclude the indices of $A_{i\mu}{}^{ab}$ can be viewed as specifying coordinates *and dimensions* within $A_{i\mu}$. We have mapped the seed fermion to a 16 × 16 = 256 dimension array. This array serves as the dimensions of the QUeST space.

1.4 Origin of Universe

We can then envision the possibility that our universe started as 1 internal dimension seed fermion and then "acquired" the 256 dimensions and 256 fundamental fermions of QUeST to form a space that includes our 3+1 space-time at the Big Bang point. This process could proceed, as it likely does, instantaneously. A corollary benefit is the location of the seed fermion in the 8-dimension UTMOST space-time.

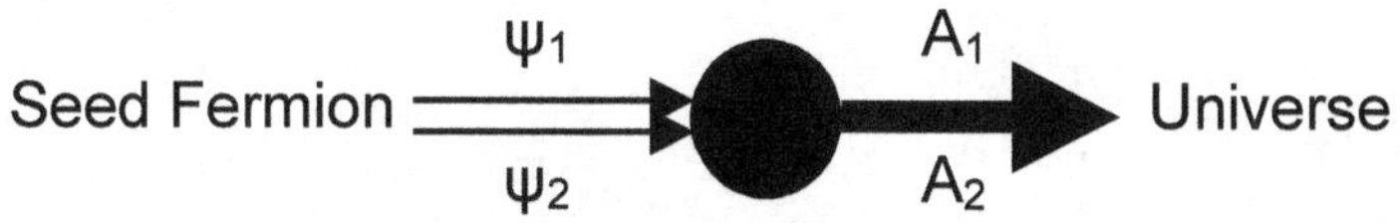

Figure 1.1. Diagram for the transition of the seed fermion to a universe particle.

Appendix 1-B shows that the Big Bang stage of the universe is analogous to the vacuum polarization of a charged particle, substantiating the interpretation of the universe as a type of particle.

Appendix 1-A. Recap of Features of QUeST Quaternion Space

This appendix describes 32 complex quaternion dimension space and some of the features of QUeST. It consists of material from Blaha (2020d) – (2020i).

1-A.1 Thirty-Two Complex Quaternion Space – 32 × 8 Dimension Array

A quaternion contains four dimensions. A complex quaternion contains eight dimensions. It is a complexification of the quaternion concept. Fig. 1-A.1 depicts the 32 dimension space. It uses a "dot" • to represent a dimension. The dimensions of the space are not assigned physically until they are mapped to internal symmetry group fundamental representation dimensions and space-time dimensions. Rather than create a cumbersome coordinate-based notation we choose to use •'s.

•••• ••••
•••• ••••
•••• ••••
•••• ••••
•••• ••••
•••• ••••
•••• ••••
•••• ••••
•••
•••• ••••

Figure 1-A.1. The 32 complex quaternion dimension QUeST array. This array is the 32 × 8 array of •'s.

The 32 × 8 form of the array is useful because it brings out the four layers of fermions that appear when the array is subdivided into four layers (8 rows each) of fundamental group representations. The Unified SuperStandard Theory[3] implied by QUeST has a matching four layers of fermions. The subdivision of Fig. 1-A.1 into layers appears in Fig. 1-A.2. The map to group representations appears in Figs. 1-A.3 and 1-A.4. We use the maps in Table 1.1 to set up the group ↔ dimension map, bearing in mind the group representations of the Standard Model:

[3] Blaha (2020d) and earlier books.

U(1)	↔	2 real dimensions
U(4)	↔	8 real dimensions
U(2)	↔	4 real dimensions
SU(3)	↔	6 real dimensions
U(1)⊗SU(2)	↔	4 real dimensions

Table 1.1. Map between fundamental representations and their dimensions.

where the dimensions have *real-valued* coordinates and are called *real dimensions*.

The eight U(4) Layer groups and eight U(4) Generation groups in Fig. 1-A.3 are present in both UST and QUeST as well as being implied by BQUeST. See Fig. 1-A.4 for the groups of one layer of the four layers.

1-A.1.1 QUeST Internal Symmetry Groups

The QUeST internal symmetry group is

$$[SU(2)\otimes U(1)\otimes SU(3)]^8\otimes U(4)^{16}\otimes U(2)^4 \qquad (1\text{-}A.1)$$

by Fig. 1-A.3 below.

1.4.1.2 QUeST Space-Time

The space-time of QUeST is 4 complex quaternion dimensions.

1-A.1.3 Generation and Layer groups of UST, QUeST and BQUeST

The Generation groups mix the fermion generations of normal and Dark sectors of each layer. The lines on the left side of Fig. 1-A.5 display Generation group mixing within each layer.

Layer groups mix fermions in all four layers for each of the four generations individually. (See right side of Fig. 1-A.5.) There are eight Layer groups: two Layer groups for Normal and Dark sectors for each generation.

The Dark groups mix between normal and Dark fermions, fermion by fermion.

1-A.1.4 Fermion- Dimension Duality

Fig. 1-A.6 shows a 1:1 relation between QUeST dimensions and the fundamental fermions of QUeST and UST. This duality is the basis of the BQUeST and BMOST one-dimension theories implying QUeST and UTMOST respectively.

1-A.1.5 Fermion Structure Extracted from QUeST Symmetry Structure

Given the form of the internal symmetries in QUeST we can determine the fermions in the fundamental group representations as shown in Fig. 1-A.5.

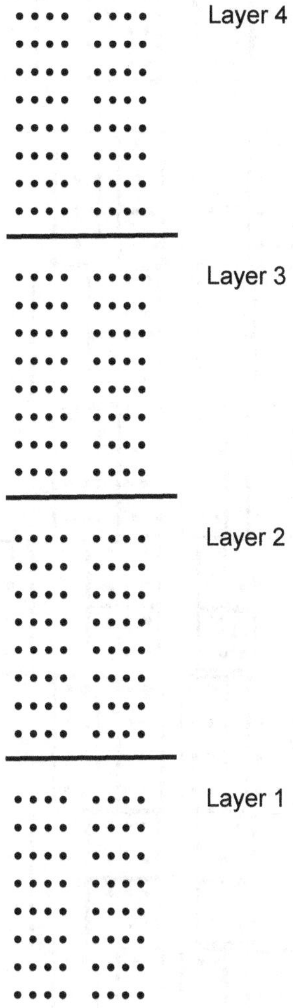

Figure 1-A.2. The 32 complex quaternion dimension QUeST array subdivided into 4 layers of 8 rows..Each layer will be seen to map to a block of fundamental group representations as shown in Figs. 1-A.3 and 1-A.4..

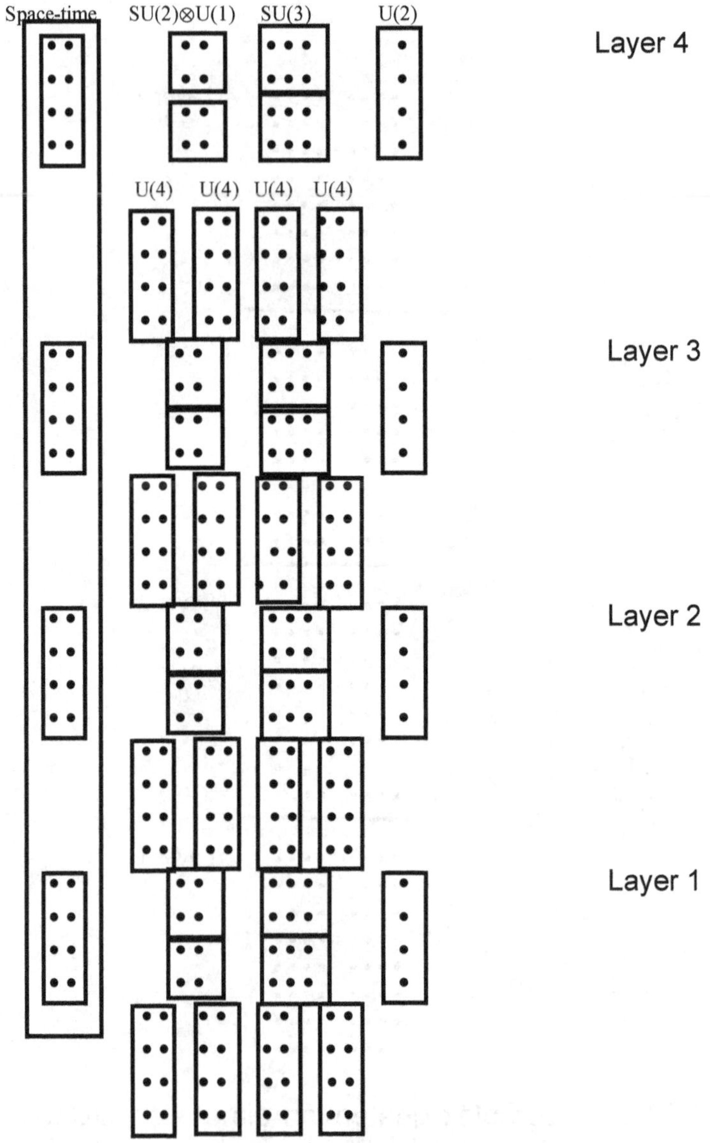

Figure 1-A.3. The four layers of QUeST internal symmetry groups (and space-time) for 32 dimension complex quaternion space. Note: each row has an 8 • complex quaternion. Note the left column of blocks combine to specify a 4 dimension complex quaternion space-time. Note each layer requires 64 dimensions.

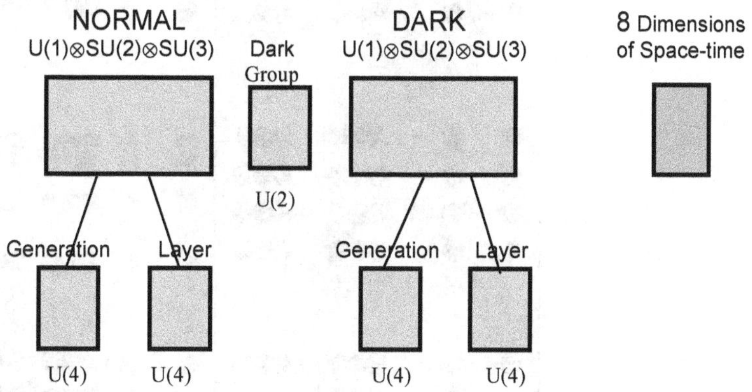

Figure 1-A.4. The internal symmetry groups for one QUeST layer of the 4 layers in the 32 × 8 dimension array format. The two large blocks are each 5 complex dimension (10 real dimension) representations of SU(2)⊗U(1)⊗SU(3). The U(2) group (badly broken) supports transformations (rotations) between Normal and Dark matter.

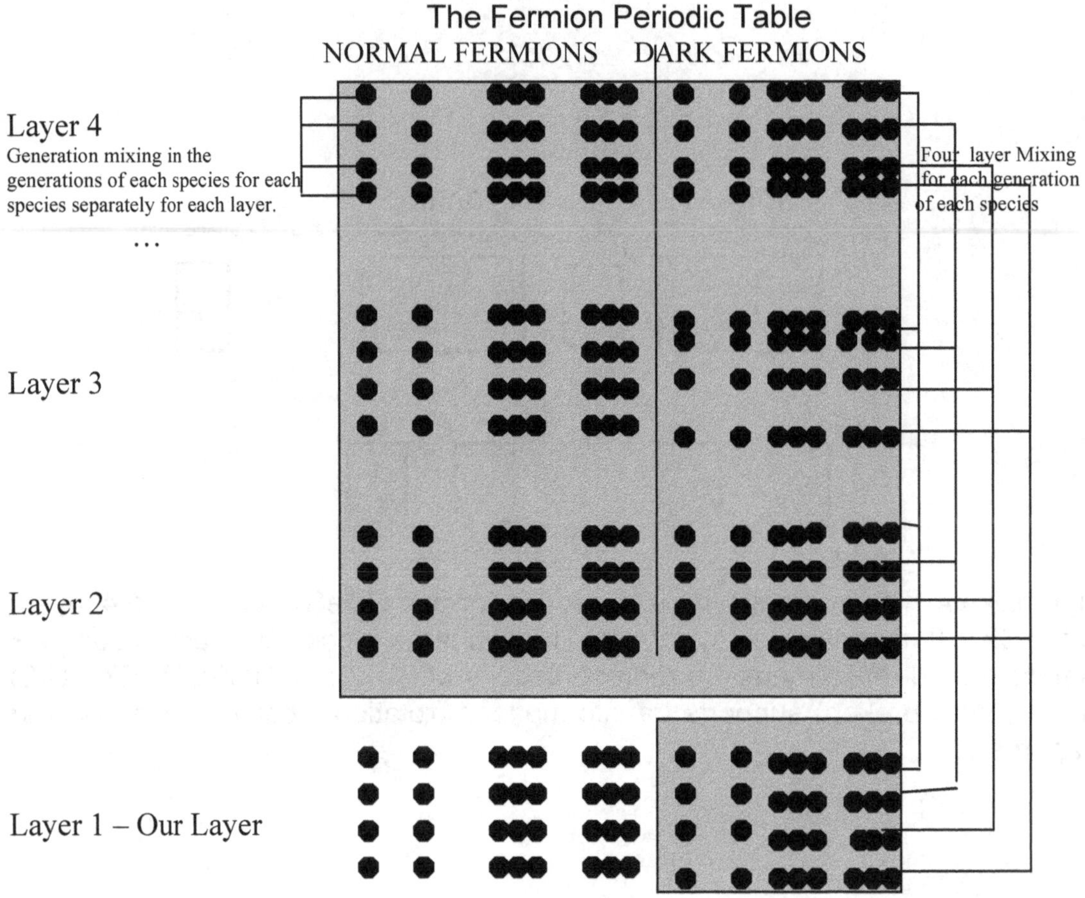

Figure 1-A.5. Fermion particle spectrum and partial examples of the pattern of mass mixing of the Generation group and of the Layer group. Unshaded parts are the known fermions with an additional, as yet not found, 4th generation. The lines on the left side (only shown for one layer) display the Generation mixing within each layer. The Generation mixing occurs within each layer using a separate Generation group for each layer. The lines on the right side show Layer group mixing (for Dark matter) with the mixing among all four layers for each of the four generations individually. There are four Layer groups for Normal matter and four Layer groups for Dark matter.. There are 256 fundamental fermions. QUeST and UST have the same fermion spectrum.

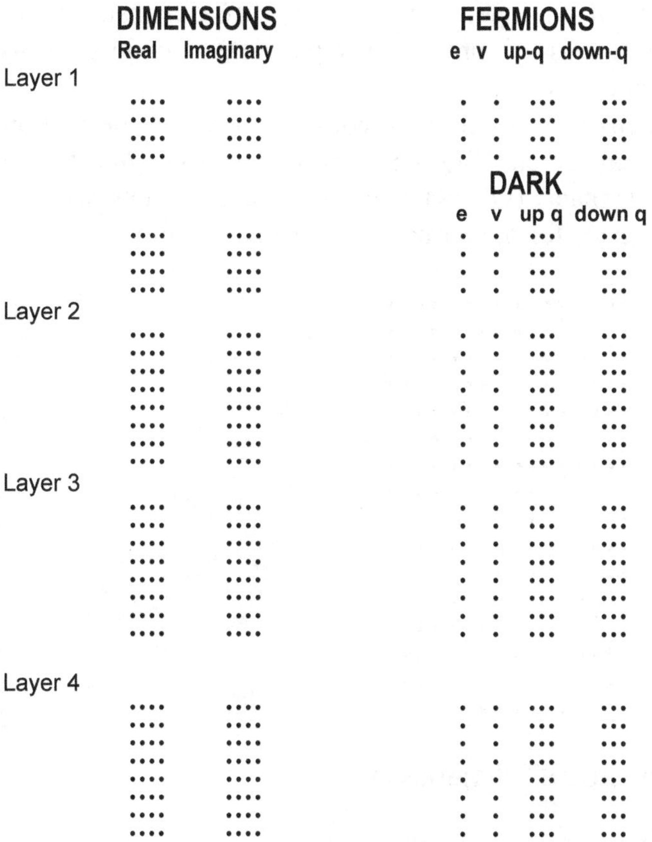

Figure 1-A.6. Fundamental fermions have a 1:1 correspondence with QUeST dimensions. Note the number of dimensions in each row is 8 – the number of dimensions in a complex quaternion. Correspondingly the number of fermions in each row is 8 – a suggestive similarity. Each layer has four normal fermion generations and four Dark fermion generations. Each dot (pebble) represents a dimension in the left part of the figure and a fermion in the right part.

1-A.2 Thirty-Two Complex Quaternion Space – 16 × 16 Dimension Array

This section describes an alternate 16 × 16 form of the QUeST dimension array. This format supports the derivation of the QUeST dimensions from BQUeST.

The 16 × 16 form of the QUeST dimension array can be based on a 16 complex octonion dimension space. The difference between this format space and the 32 dimension space is not physically meaningful at present. The difference will be physically meaningful if the masses of the fermion spectrum and the full pattern of symmetry breaking is determined. Then one can differentiate between a four layer theory as above in section 1-A.1 and a two "layer" theory presented below in this section.

The alternate 16 × 16 form of the QUeST dimension array is simply constructed from the preceding 32 × 8 dimension array by "moving" the 16 × 8 lower half of the 32

× 8 array of Fig. 1-A.1 to the "right" of its upper half. The 16 × 16 dimension array appears in Fig. 1-A.7. Note that the forms are physically equivalent if done prior to mapping dimensions to group representations.

The 16 × 16 form of the dimension array is more convenient for our derivation of a fundamental basis for QUeST that we call BQUeST (pronounced bee-quest). No change in the number of dimensions is made. (QUeST could also be viewed as residing in a 16 complex octonion dimension space for the purpose of dimension counting.)

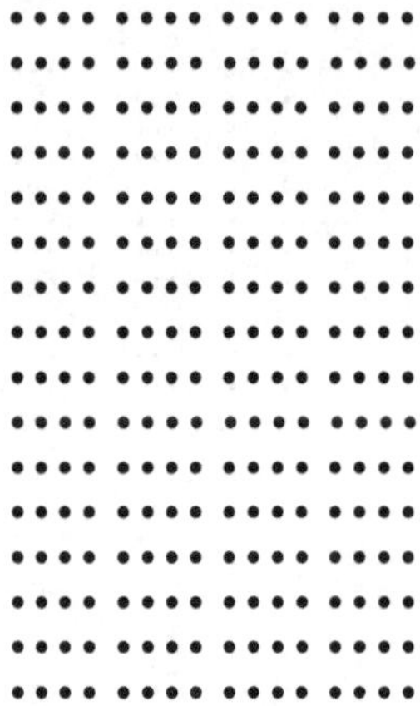

Figure 1-A.7. The 16 × 16 array of QUeST dimensions.

Fig. 1-A.8 shows the assignment of fundamental group representations to the 16 × 16 dimension array. Note this form of array has two "layers" of 128 dimensions in contrast to the four 64 dimension layers of Fig. 1-A.3. The difference can be attributed to the third and fourth layers of Fig. 1-A.3 becoming "extra" Dark sectors in "layers" 1 and 2.

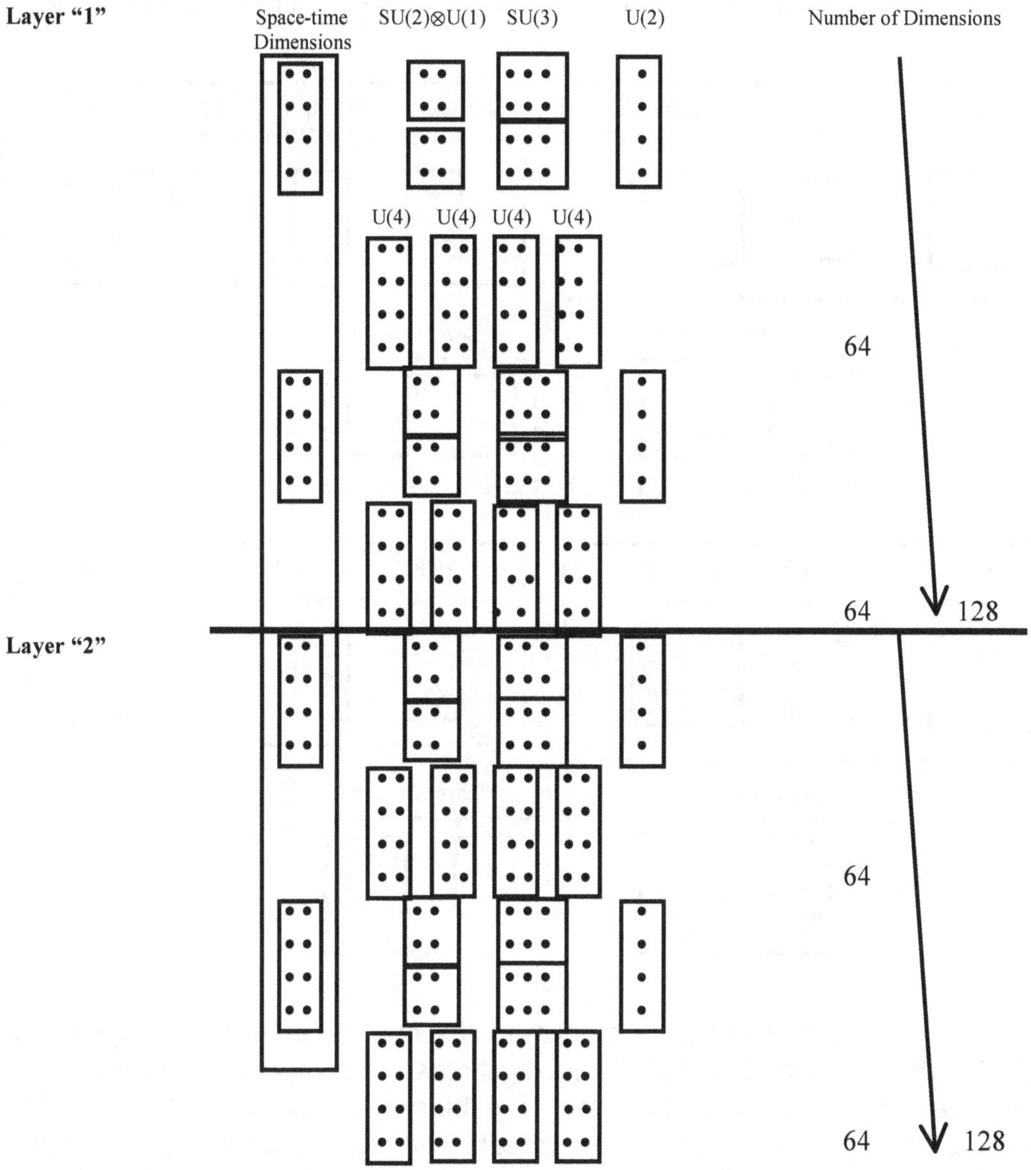

Figure 1-A.8 The 16 × 16 form of QUeST array has two "layers", each of which is composed of two layers of the 32 × 16 four layer figure of Fig. 1-A.3. Note each of these "layers" has 128 dimensions.

The group representations in Fig. 1-A.8 can be subdivided into blocks of 16 dimensions as shown in Fig. 1-A.9 and 1-A.10 for the "layers" of Fig. 1-A.8. Blaha (2020i) provides a rationale for 16 dimension subblocks based on U(8) considerations.

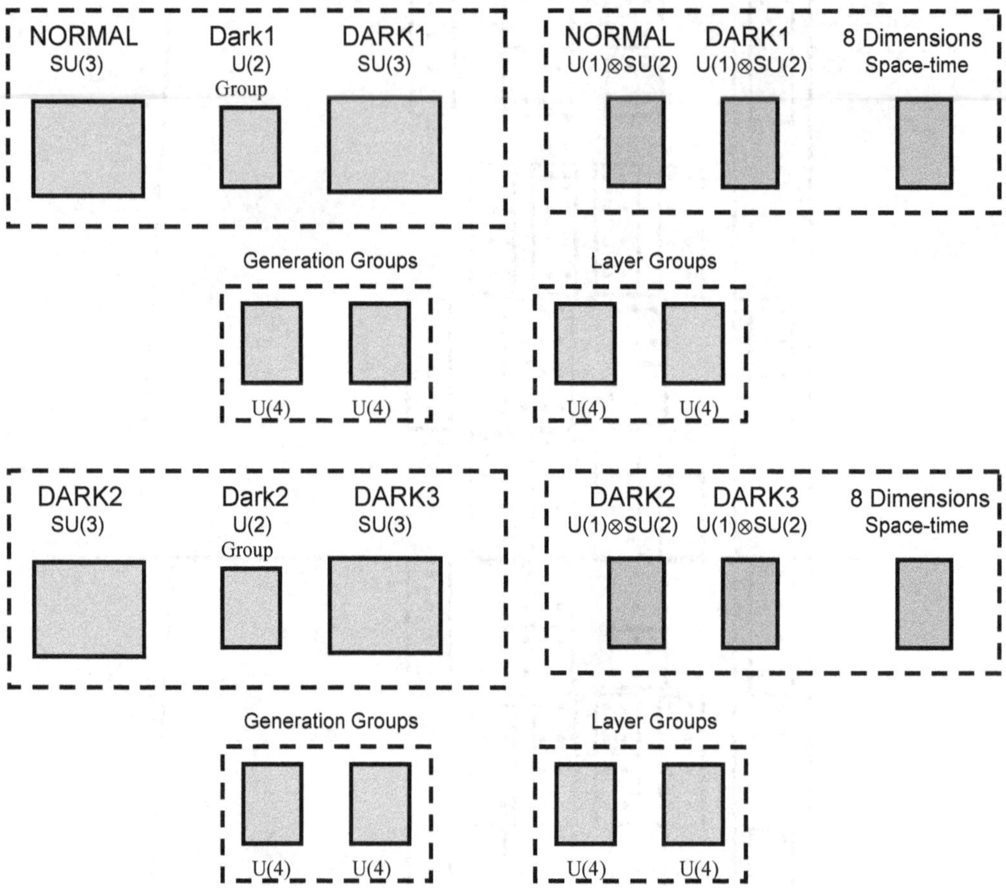

Figure 1-A.9. Set of 16 dimension blocks for the first "layer" (of two "layers") of the 16 × 16 array. Each "dashed" block (regardless of its apparent size) is a 4 × 4 = 16 array of dimensions. This set of 8 blocks contains the 8×16 = 128 dimensions of "layer" 1. "Layer" 2 is similar. The Dark U(2) groups supports transformations (rotations) between the types of matter: Normal and Dark1; and Dark2 and Dark3.

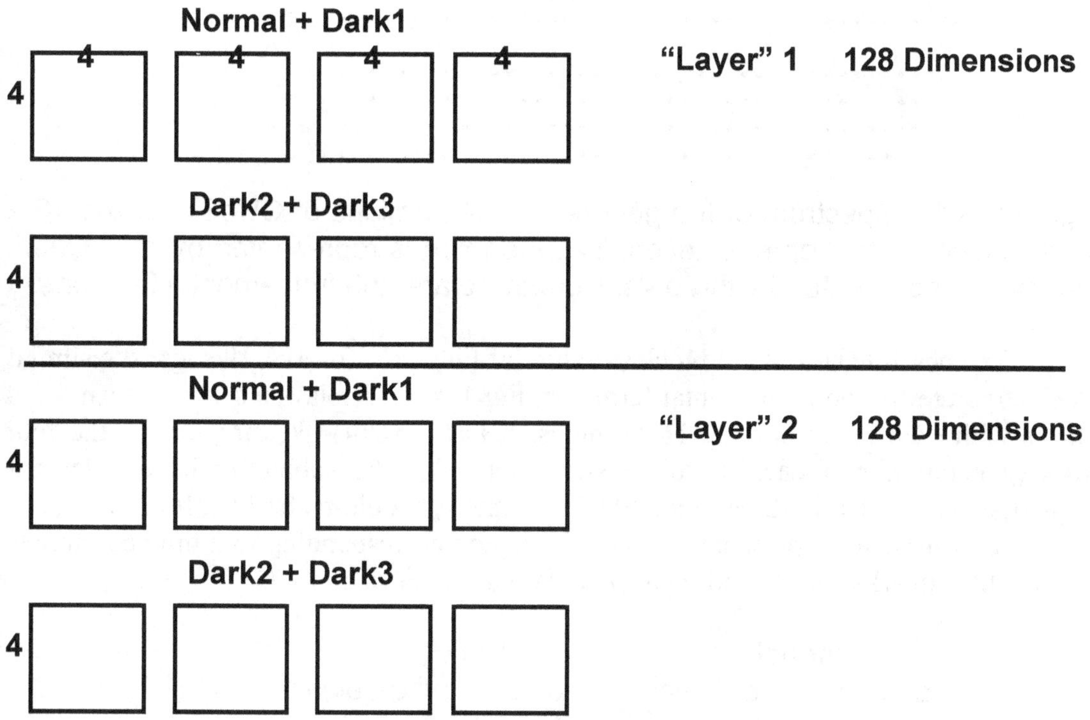

Figure 1-A.10. The two "layers" of 4 × 4 dimension subblocks of the 16 × 16 dimension array. .

QUeST fermions have a similar format to the dimension subblock structure in Fig. 1-A.10. Fig. 1-A.11 displays the 256 fermions of QUeST. The shift from a 32 × 8 dimension array with four layers of groups to a 16 × 16 dimension array gives a two "layer" form with the" lower" two layers of the 32 × 8 dimension array becoming the Dark2 and Dark3 parts of the two "layer" form. This shift has no discernable physical consequences as far as our considerations are concerned since we are moving Dark sectors. When mass generation and symmetry breaking are better understood, a physically significant difference may be evident.

Figure 1-A.11. Spectrum of the generations of fermions of QUeST for the 16 × 16 dimension array representation. Each fermion is represented by a •. Quark triplets are represented by three •'s. Note there are 256 fundamental fermions.

The possible U(8) 4 × 4 block structure of Figs. 1-A.9 – 1-A.10 suggest a similar block structure for the fundamental fermions. Fig 1-A.12 displays a set of sixteen 4 × 4 blocks with each block holding 16 fermions. It has a SU(4)-like structure of the four rows (generations) of each 16 dimension subblock. The subblocks have a lepton-triquark content. Clearly the implied SU(4) symmetry would be broken. It is suggestive of a Lorentz 4-vector representation with the lepton corresponding to a time coordinate and the three quarks corresponding to spatial coordinates.

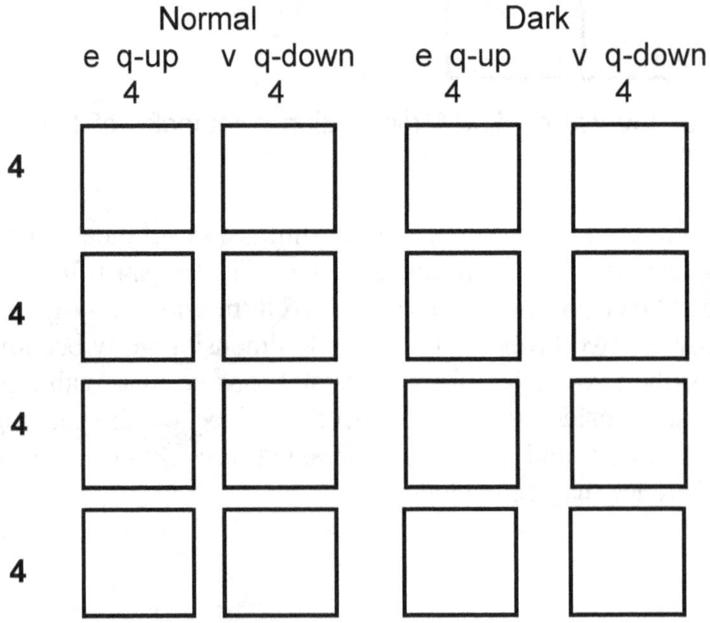

Figure 1-A.12. Block form of a 16 × 16 QUeST fermion array with each block row corresponding to one layer. Each block contains four generations of fermions. The result is 4 × 4 blocks. The label e q-up indicates a charged lepton

– up-type quark pair, v q-down indicates a neutral lepton – down-type quark pair, and so on. Note the blocks can be reaaranged into a 32 × 8 form without physical consequences at this level of discussion since the right two columns and the lowest two rows are all Dark at present.

1-A.3 A Partition to Real 3+1 Dimension Space-times

We can partition Fig. 1-A.4 into real and imaginary subspaces with 3+1 dimension space-times.

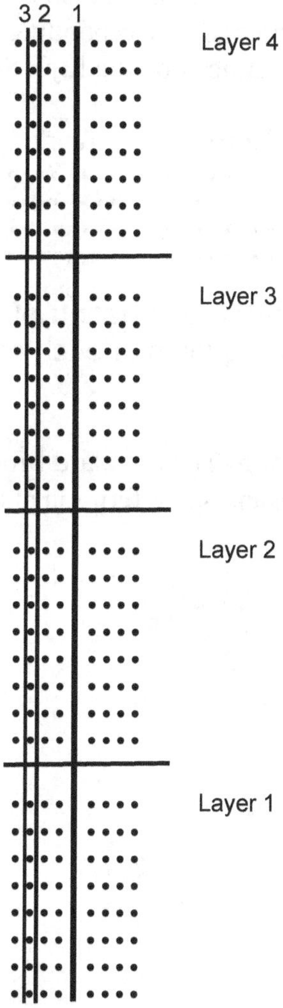

Figure 1-A.13. The 32 complex quaternion dimension QUeST array partitioned down to a real spwce-time. The partition labeled "1" reduces the array to 32 quaternion dimensions after discarding the right columns. The partition labeled "2" reduces the array to 32 complex dimensions similarly. The partition labeled "3" reduces the array to 32 real-valued dimensions in the leftmost column..

The real-valued column of dimensions numbers 32 dimensions. It maps to the internal symmetry group:

$$[SU(2) \otimes U(1) \otimes SU(3)]^2 \otimes U(4) \qquad (1\text{-A}.2)$$

where the U(4) group is the Generation group, which implies 4 generations. There is an SU(2)⊗U(1)⊗SU(3) group giving the ElectroWeak and Strong interactions of Normal matter. The other SU(2)⊗U(1)⊗SU(3) group gives the Dark Matter ElectroWeak and Strong Interactions. Note the Layer group is not present.

These groups require 28 dimensions. The remaining four dimensions give 4 dimension space-time. The total is 32 dimensions.

The fundamental fermions corresponding to the internal symmetry groups number 32 fermions (fermion-dimension duality). See Fig. 1-A.14.

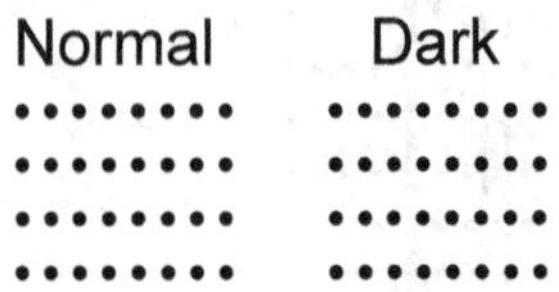

Figure 1-A.14. Spectrum of the generation of fermions. Note only one layer of fermions. Two leptons and 6 quarks in each row of Normal matter. Similarly for Dark matter.

This restricted model is similar to The Standard Model: same internal symmetry, same 4 generations, same space-time, and same fermion and vector boson spectrums..

Appendix 1-B. Evidence for Universe Particles

This appendix provides evidence for the existence of a Megaverse and for the possibility that universes are particles. It is abstracted from Blaha (2018e), which provides much more detailed data.

1-B.1 Evidence for Entities Beyond the Universe

At first glance it would seem impossible to produce evidence for the existence of other universes. However there are subtle means by which we can 'sense' experimentally 'nearby' universes should they exist. The mechanism would appear to be gravitational effects exerted on objects within our universe by unseen objects of enormous mass. Currently there appears to be three experimental suggestions of the existence of 'nearby' universes and one theoretical argument based on an influx of mass-energy from the Megaverse that may cause the expansion of our universe.

There are also theoretical motivations for believing that there are entities beyond our universe. These are detailed in chapter 30 of Blaha (2018e).

1-B.1.1 Great Attractors

One potential support is the discovery of the Great Attractor (at the center of the Laniakea Galaxy Supercluster), and the more massive Shapley Attractor (centered in the Shapley Supercluster)[4]. These attractors contain massive numbers of galaxies and are drawing galaxies over a distance of millions of light years towards them.

If another universe(s) is 'near' our universe it could act as a 'gravitational magnet' and draw galaxies within our universe towards it to form one or more superclusters which could then act as attractors. Thus attractors might indirectly reveal the presence of other nearby universes—contrary to the expected large scale uniformity of the universe. The only other apparent source of superclusters is chance. Chance seems an unsatisfactory possibility in the present case.

1-B.1.2 Bright Bumps in Universe Suggesting Collision with Another Universe

A recent study[5] of the residual brightness of parts of the accessible universe found that bright patches appeared if a model of the CMB (Cosmic Microwave Background) with gases, stars and dust was 'subtracted' from the PLANCK map of the entire sky. After the subtraction one would expect only noise spread throughout the sky. However, bright patches were seen in a certain range of frequencies. These anomalies are thought to be a result of our universe colliding with another object – presumably another universe in the Megaverse.

---

[4] Tully, R. Brent; Courtois, Helene; Hoffman, Yehuda; Pomarède, Daniel, "The Laniakea Supercluster of galaxies". Nature (4 September 2014). 513 (7516): 71–73; arXiv:1409.0880.

[5] Ranga-Ram Chary, arXiv.org:/1510.00126 (2015).

1-B.1.3 Cold Spot in Universe Suggesting Collision with Another Universe

Another recent study[6] of a huge cold region of the universe spanning billions of light years revealed that this region is not a relatively empty region but rather is similar to in its distribution of galaxies to the rest of the universe. Previous the Cold Spot (an area where cosmic microwave background radiation – the leftover Big Bang radiation is weak – making it significantly colder (0.00015C colder) than the average temperature of the universe.)

An analysis of 7,000 galaxy redshifts using new high-resolution data has now shown that the Cold Spot is similar to the rest of the universe. The Durham University group suggested that the Cold Spot might have been caused by a collision between our universe and another Universe. They further suggested that there is only a 1 in 50 chance that it could be explained by standard cosmology.

Thus we have another important piece of circumstantial evidence in favor of other universes and thus the Megaverse.

1-B.1.4 Megaverse Energy-Matter Infusion into Our Universe

In chapter 14 of Blaha (2017c) we presented a model for an influx of mass-energy from the Megaverse to support the Bondi-Gold-Hoyle-Narlikar Steady State Cosmology, which was originally based on the 'continuous creation of mass-energy' by Hoyle and Narliker. This model explains why the value of Ω makes the universe close to flat. If this model is correct then we would have concrete support for a Megaverse with a low mass-energy density leaking mass-energy into our universe. *More generally, it suggests that universes are surfaces of high mass-energy density in a Megaverse of low mass-energy density – with a ratio of mass-energy densities of the other of 10^{30}.*

1-B.1.5 Conclusion

We conclude that data is beginning to emerge favoring multiple universes and a physical Megaverse in support of the theoretical justifications presented earlier.[7]

1-B.2 Hubble Constant and Universe Expansion

Our universe is clearly expanding from an initial state called the Big Bang to its current state. The Hubble "Constant" measures the rate of growth. Much of the expansion data on the Hubble Constant is presented below. After reviewing the data we propose a fit to the data in this chapter that explains the apparent growing rate of expansion. Below we derive the form of the fit.

1-B.2.1 Hubble Constant Experimental Data

There are a number of astrophysical studies of the universe that suggest that the Hubble Constant is *not* constant. Although there are significant margins of error it appears that the early universe "beginning" epoch around 380,000 years had a Hubble Constant of 67.8 km s^{-1} Mpc^{-1}.[8] More recently, red shift studies of quasars have given a

---

[6] T. Shanks et al, Durham University (Australia), Monthly Notices of the Royal Astronomical Society, 2016 .
[7] Chapter 59 of Blaha (2020c).
[8] See, for example, K. Aylor *et al*, arXiv:1811.00537v1 (2018) based on studies of the cosmological sound horizon.

Hubble Constant of 73.2 km s^{-1} Mpc^{-1}.[9] And studies of binary black hole merger gravity waves[10] have given a Hubble Constant of 75.2 km s^{-1} Mpc^{-1} (and earlier of 78 km s^{-1} Mpc^{-1}). Another study of events at 1.8 billion ly yielded a Hubble Constant of 70.0 km s^{-1} Mpc^{-1}.[11] Further studies have given the Hubble Constants: 1) Of variable stars 73.2 km s^{-1} Mpc^{-1}, 2) Of light bent by distant galaxies 72.5 km s^{-1} Mpc^{-1}, 3) Of Magellan Cepheids 74.03 ± 1.42 km s^{-1} Mpc^{-1}, [12] 4) Of distant red giant[13] brightness 69.8 km s^{-1} Mpc^{-1},

The only apparent conclusion at this time is that there was a Hubble Constant (Constant) H of approximately 67.8 km s^{-1} Mpc^{-1} early in the universe, and ranging up to 75.2 km s^{-1} Mpc^{-1} at the current time. Thus an increasing Hubble Constant.

For the purpose of discussing the apparent increase in H with time, we average the above eight "recent" values of H in the spirit of Bayesian equal probability to obtain a **recent time Hubble average of 73.24** km s^{-1} Mpc^{-1}.[14] Thus there appears to be a 7% - 9% increase in the Hubble Constant over time.

1-B.2.2 Fit to the Hubble Constant Data and Scale Factor

It is generally expected that the Hubble Constant will decline with time from the time of the Big Bang. It is generally believed that the Hubble Constant has recently been increasing with time. **The declining value in the past and the current growth of the Hubble Constant imply that it reached a minimum at some time in the past.**

Our fit to the data from Blaha (2019c) and (2019e) was

$$\begin{aligned} a(t) &= (t/t_{now})^{g + ht} \\ &= \exp[(g + ht)\ln(t/t_{now})] \end{aligned} \tag{1-B.1}$$

where g and h are constants. (The constant h is *not* the Hubble parameter.) There is an "ht" term in the exponent based on the rise in H(t) suggested by experimental data.

Eq. 1-B.1 can be approximated by

$$a(t) = (H_0 t)^{g + H_0 t} \tag{1-B.2}$$

which nicely relates it to the Hubble parameter H_0.

The basis of the fit for a(t) was:

1. Power law behavior (in part) as in the radiation and matter dominated approximations.

---

[9] M. Soares-Santos *et al* , arXiv:1901.01540 (2019).
[10] DES and LIGO collaborations *et al*, arXiv:1901.01540 (2019).
[11] B.P. Abbott *et al*, arXiv:1710.05835 (2017).
[12] J. T. Nielsen *et al*, Marginal evidence for cosmic acceleration from Type Ia supernovae, Nature Scientific Reports (2016); arXiv:1506.01354 (2015). A. Riess *et al*, The Astrophysical Journal **875**, 145 (2019) and references therein. A. Riess *et al*, arXiv:1903.07603 (2019).
[13] W. Freedman *et al*, The Astrophysical Journal **880** (July, 2019).
[14] In Blaha (2019c) and (2019e) we used an average estimate of 73.7 km s^{-1} Mpc^{-1}.

2. The known shape of H(t) at early times, and at present, as described above

3. The simplicity of the fit. Two values of H(t) set the constants g and h.

4. Faster than exponential future growth with no Big Rip.

The Hubble Constant implied by eq. 1-B.1 is

$$H(t) = (da/dt)/a = g/t + h(1 + \ln(t/t_{now}) \qquad (1\text{-B}.3)$$

We set the value of H(t) by using its value at two values of time determining g and h. Based on experimental data:

$$H(t_c) \equiv H(380{,}000\ \text{yr}) = 67.8\ \text{km s}^{-1}\ \text{Mpc}^{-1} \qquad (1\text{-B},4)$$
$$H(t_{now}) = 73.24\ \text{km s}^{-1}\ \text{Mpc}^{-1}$$

and

$$h = (t_c H(t_c) - t_{now}H(t_{now}))[\ t_c - t_{now} + t_c \ln((t_c/t_{now})]^{-1} \qquad (1\text{-B}.5)$$
$$g = (H(t_{now}) - h)\ t_{now}$$

where t_c = 380,000 years after the Big Bang.[15] We obtained

$$h = 2.25983 \times 10^{-18}\ \text{s}^{-1} = 1.49 \times 10^{-33}\ \text{eV} \qquad (1\text{-B}.6)$$
$$g = 0.000282377 = 2.82377 \times 10^{-4}$$

There are two approaches to the universal scale factor fit of eq. 59.1. One approach is based on a remarkable coincidence between the power g in the fit and the QED power g seen earlier. It leads to a theory in which the expansion of the universe taken over all time is a vacuum polarization phenomenon. The other approach is based on the Einstein equation for the scale factor. We show that the Universal Scale Factor is consistent with the Einstein equation if additional (dark) energy is properly taken into account.

1-B.3 Hubble Parameter and Vacuum Polarization of a Particle

In this section we will show that the initial behavior of the expanding universe's scale factor is the same as the exponent of the vacuum polarization of a particle as its energy gets very large.

1-B.3.1 Vacuum Polarization Generation of the Early Time Part of the Universal Scale Factor

Perhaps the crowning achievement of our universal scale factor eigenvalue formulation[16] for coupling constants is the successful relation of universe evolution to vacuum polarization due to a vector QED-like interaction between universes.

---

[15] Based on the data value of 67.8 km s^{-1} Mpc^{-1} at t = 380,000 years.

[16] See chapter qqzz.

In massless QED we found that the vacuum polarization had the form:[17]

$$F_1(\alpha)(p/\Lambda)^{2g_{QED}} \quad (1\text{-B.7})$$

where $F_1(\alpha)$ is the "eigenvalue function" for the Fine Structure Constant[18] of the Johnson-Baker-Willey model of massless QED, p is the momentum, and Λ is the ultraviolet cutoff. The value of g_{QED} that corresponded to the Fine Structure Constant is

$$g_{QED} = -\,0.00058053691948 \quad (1\text{-B.8})$$

and the Fine Structure Constant was correctly found to be

$$\alpha_{calculated}(g_{QED}) = 0.0072973525693 \quad (1\text{-B.9})$$

to 13 digit accuracy according to the *Particle Data Table of 2019.*

Comparing our Universal Scale Factor g value (eq. 1-B.6) ,which governs early time behavior of the expanding universe, with g_{QED} we find

$$-g = 0.000282377 \cong -\tfrac{1}{2}g_{QED} = -0.000290268 \quad (1\text{-B.10})$$

1-B.3.2 Comparison of QED Vacuum Polarization Exponent with Universe Vacuum Polarization Exponent

Eq. 1-B.10 shows the numeric values of the g powers are approximately equal up to a factor of -2. The QED exponent describes high energy vacuum polarization behavior. The universe power g describes the small time universe expansion (near the Big Bang). The relation between the values of g and g_{QED} clearly suggests an analogy.

Further the low energy (infrared) behavior of the QED vacuum polarization which is mass dependent is analogous to the large time (recent time) behavior of a(t) which is governed by the h term in the exponent of a(t).

We now show the universe vacuum polarization is due to a new vector interaction between universes, and show that it is related to the QED vacuum polarization by eq. 1-B.10.[19]

1-B.3.3 A New Vector Interaction for Universe Particles

We assume universes can be treated as particles in 4-dimensional space-time.[20] Since experiments appear to have shown that our universe does not rotate (does not

---

[17] Eq. 12 in S. Blaha, Phys Rev **D9**, 2246 (1973).
[18] The author calculated $\alpha = 1/137\ldots$ exactly in Blaha (2019a) and (2019b).
[19] The following subsections appeared in Blaha (2019c).
[20] Universes are composite entities but we can treat them as quantum particles in the same manner as physicists treated protons and neutrons etc. as quantum particles before quark theory was accepted. See Blaha (2018e) for a detailed discussion of universe particles.

have spin)[21] we will assume the universe is a spin 0 boson. We assume that universes have a vector field interaction similar to QED.

Given this QED-like framework, then universe-antiuniverse pair production and vacuum polarization becomes possible. We assume the QED-like boson lagrangian

$$\mathcal{L} = \tfrac{1}{2}\,(\partial_\mu\varphi^\dagger\partial^\mu\varphi - m^2\varphi^\dagger\varphi) - ie_0{:}\,\varphi^\dagger(\overrightarrow{\partial_\mu} - \overleftarrow{\partial_\mu})\,\varphi{:}\,A^\mu + e_0^2{:}A^2{:}\,{:}\varphi^\dagger\varphi{:} + \delta m^2{:}\varphi^\dagger\varphi{:} \tag{1-B.11}$$

where $\varphi(x)$ is a "charged" quantum universe scalar particle field[22] and A^μ is a QED-like field. We now proceed to calculate the second order vacuum polarization of a universe particle. We will assume the term in $\mathcal{L}$ linear in A^μ is the relevant term since the quadratic term always is negligible compared to the linear term in each order α^n of perturbation theory by a factor of α. The neglected terms will be assumed to not affect the calculated eigenvalue function.

1-B.3.4 Second Order Vacuum Polarization of a Scalar Universe Particle

The one loop vacuum polarization Feynman diagram is

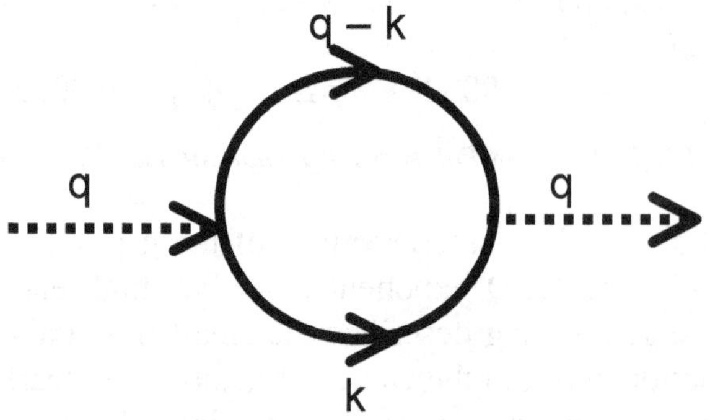

Figure 1-B.1 One loop vacuum polarization boson Feynman diagram.

Its evaluation is

$$I_{\mu\nu} = (-ie_0)^2 \int \frac{d^4k}{(2\pi)^4}\,\frac{i}{(k^2 - m^2 + i\varepsilon)}\,\frac{i}{(k^2 - m^2 + i\varepsilon)}\,(q-2k)_\mu(q-2k)_\nu \tag{1-B.12}$$

$$= \frac{\alpha}{2\pi}\int_0^\infty dz_1 \int_0^\infty dz_2\, \frac{g_{\mu\nu}\exp[i(q^2 z_1 z_2/(z_1+z_2) - (m^2+i\varepsilon)(z_1+z_2))]}{(z_1+z_2)^3} + \text{gauge terms}$$

upon introducing parameters z_1 and z_2 to enable exponentiation and integration over k, where

[21] The lack of universe rotation (spin) is indicated by a study of Cosmic Microwave Background (CMB) by D. Saadeh *et al*, Phys. Rev. Lett. **117**, 313302 (2016).

[22] The charge is not electromagnetic charge.

$$\alpha = e_0^2/4\pi \qquad (1\text{-B}.13)$$

After some steps we found

$$I_{\mu\nu} = \frac{i\,\alpha}{12\pi} q^2 g_{\mu\nu} \ln(\Lambda^2/m^2) + \ldots \qquad (1\text{-B}.14)$$

with finite and other gauge terms not shown.

Thus we find the renormalization constant Z_{3U} for a scalar universe particle is

$$Z_{3U} = 1 - \alpha/12\pi \ln(\Lambda^2/m^2) \qquad (1\text{-B}.15)$$

If we let

$$\alpha_U = \alpha/4 \qquad (1\text{-B}.16)$$

then we obtain the form similar to the one loop value of Z_3 for spin ½ electron QED:

$$Z_{3U} \cong 1 - \alpha_U/3\pi \ln(\Lambda^2/m^2) \qquad (1\text{-B}.17)$$

We now *provisionally assume* that α is the QED fine structure constant. We denote it as α_{QED}. We verify this choice later.

Thus the "fine structure constant" α_U for our vector interaction is

$$\alpha_U \equiv \alpha_{QED}/4 = 0.001824338 \qquad (1\text{-B}.18)$$

We now turn to the Johnson-Baker-Willey (JBW) model of massless QED since at ultra-high energy our vector interaction theory with lagrangian eq. 1-B.11 becomes the JBW model for a scalar particle. In the JBW model we calculated α_{QED} and found the corresponding power of the Z_3 divergent factor which we denote g_{QED}.

1-B.3.5 Finding the Universe g_U

Now we perform the same calculation for universe vacuum polarization and find the g value, which we denote g_U, corresponds to α_U. The value of g_U will be seen to lead to the power g in the universal scale factor almost exactly.

The universe eigenvalue function is[23]

$$F_2(\alpha_U) = F_1(\alpha_U) - [2/3 + \alpha_U/(2\pi) - (1/4)[\alpha_U/(2\pi)]^2] \qquad (1\text{-B}.19)$$

For

$$\alpha_U \equiv \alpha_{QED}/4 = 0.001824338 \qquad (1\text{-B}.20)$$

we found the eigenfunction value

$$F_2(\alpha_U = 0.001824338) = 5.10824 \times 10^{-12} \cong 0 \qquad (1\text{-B}.21)$$

---

[23] We assume the universe eigenvalue function has the same form as the QED eigenvalue function.

Examining $F_2(\alpha_U)$[24] as a function of g_U we found the value of g_U corresponding to α_U is

$$g_U = -0.00014525 \qquad (1\text{-B.22})$$

Thus the universe vacuum polarization is

$$\Gamma_U(p) = (p/\Lambda)^{2g_U} \qquad (1\text{-B.23})$$

The fourier transform is[25]

$$a(t) = (1/2\pi)\int_0^\infty dp/p\, \exp(-ipt)\, \Gamma_U(p) \qquad (1\text{-B.24})$$

$$= k\,(t/T)^{-2g_U} \qquad (1\text{-B.25})$$

where k is a constant and where

$$1/T = \Lambda \qquad (1\text{-B.26})$$

with Λ being the "momentum space" cutoff mass. Comparing to 1-B.10 we find

$$g = -2g_U$$
$$= 0.0002905 \qquad (1\text{-B.27})$$

From eq. 1-B.25 for the power g of a(t) we see the universal scale factor g is

$$g = 0.000282377 \qquad (1\text{-B.28})$$

Thus the value of g calculated from the universe vacuum polarization differs from the actual value of g by less than 3%. Given the approximate nature of our JBW calculation of vacuum polarization the agreement is remarkable.[26]

In addition we found the "fine structure constant" for the vector interaction to be given by eq. 1-B.16 resulting in

$$e_U = (4\pi\alpha_U)^{1/2} = 0.151411 \qquad (1\text{-B.29})$$

Thus we have shown the universe vacuum polarization $\Gamma_U(p)$ when transformed to time is the universal scale factor a(t) up to a constant. The evolution

---

[24] $F_2(\alpha_U)$ and $F_2(g_U)$ are alternate notations for the same function.

[25] Those who might object to fourier transforming to time t should remember that inside a Black Hole the "time-like" coordinate is the radius and the time variable t is comparable to a spatial coordinate. The possibility that the universe is a Black Hole is not excluded. This fourier transform appears in Blaha (2019c) in eq. 25.25 with a typographic error—the division by p was omitted.

[26] And may be exact! The value of the Hubble Constant H in recent times varies from about 70 – 75 making the calculation of g also approximate. We chose an average value of 73.24 to obtain the value of g above. If we chose the current value for H to be 75.58 we would have $g = -2g_U$ exactly. Note: studies of binary black hole merger gravity waves have given a Hubble Constant of 75.2 km s^{-1} Mpc^{-1} (and earlier of 78 km s^{-1} Mpc^{-1}), and studies of light bent by distant galaxies give H = 72.5 km s^{-1} Mpc^{-1}. Thus the value H = 75.58 is not unreasonable.

of our universe is set by universe vacuum polarization. Other 4D universes may be expected to be similar.

The above relation we have found between QED-like vacuum polarization and universe vacuum polarization (Dark Energy) appears to confirm our interpretation of universe Dark Energy as mainly a consequence of universe vacuum polarization due to a universe vector interaction.[27]

1-B.3.6 Dark Energy is Equivalent to Universe Vacuum Polarization

Dark Energy is elusive both on the experimental and theoretical levels. We know it exists through its effects on our universe. Yet interactions with matter have not been found. Thus it is somewhat of a phantom.

The existence of Dark Energy, which, clearly, strongly affects the evolution of the universe, means that the Einstein equation, usually regarded as central to universe evolution, is incomplete for that purpose. It does not specify the total energy density ρ_{tot}.

$$\dot{a}^2 - 8\pi G\rho_{tot}a^2/3 = -k \qquad (1\text{-B.30})$$

However we can obtain a "handle" on the total energy density by inserting our universal scale factor a(t) in the Einstein equation together with the known radiation density, matter density and Cosmological Constant Λ terms:

$$\rho_{tot}(t) \equiv \rho_{crit}\Omega_{tot}(t) = \rho_{crit}[\Omega_\Gamma(t) + \Omega_M(t) + \Omega_\Lambda + \Omega_T]$$

where the unknown part needed to makes the Einstein equation correct is the elusive Dark Energy $\rho_T(t)$

$$\rho_T(t) = \rho_{crit}\Omega_T(t) \qquad (1\text{-B.31})$$

1-B.3.7 Quasi-Free Universe Particles

Since $F_2 \cong 0$, universe particles are very much like free particles. (The vacuum polarization is effectively zero.

Universe particles are not totally free particles due to gravitation and Standard Model interactions such as electromagnetism. We treated the case of free universe particles in Blaha (2018e).

1-B.3.8 Doubling Relation Between Coupling Constants

The coupling constants that we have derived show a doubling whose fundamental significance remains to be understood.

---

[27] Rather like the discovery of the Ω^- particle in the 1960s confirmed Gell-Mann's SU(3) theory.

INTERACTION	COUPLING CONSTANT[28]
Universe Interaction e_U	0.1514
QED $e_{QED} = (4\pi\alpha_{QED})^{1/2}$	0.303
Weak SU(2) g_W	0.619
Strong SU(3) g_S	1. 22

Figure 1-B..2 The interaction coupling constants show a regular doubling. A fundamental cause for doubling is not apparent.

[28] M. Tanabashi *et al* (Particle Data Group), Phys. Rev. D**98**, 030001 (2018).

Appendix 1-C. PseudoQuantum Field Theory

This Appendix describes PseudoQuantum field Theory. We begin by describing advantages of PseudoQuantum Field Theory (from Blaha (2020c) and earlier books). To these advantages we add the capability of supporting the derivation of QUeST from BQUeST seen earlier, and supporting the derivation of UTMOST from BMOST that will be seen later. The advantages are:

1. Quantization in any coordinate system in flat or curved space-times with an invariant definition of asymptotic particle states. An n particle asymptotic state in one coordinate system is a unitarily equivalent n particle asymptotic state in any other coordinate system. Therefore particle number is invariant under change of coordinate system. This is important for the Unified SuperStandard Theory in curved space-times. It is also important for quantization in higher dimensional Euclidean spaces such as the Megaverse. The method was developed in the late 1970's by the author to provide a quantization procedure which supports a unique particle interpretation of states in arbitrary non-static space-times where no global timelike coordinate (Killing vector) exists. PseudoQuantum Field Theory which we developed in a series of books[29] also can be formulated in the Megaverse. Thus we can use it in the Megaverse to implement the Higgs Mechanism to generate particle masses and symmetry breaking.

2. PseudoQuantum Field Theory enables one to define Higgs particle dynamics in such a way that a non-zero vacuum expectation value cleanly separates from the quantum field part of the Higgs fields. This technique can be used in symmetry breaking mechanisms, mass generation, and possible generation of coupling constants as vacuum expectation values.

3. It supports the canonical definition of higher derivative field theories through the use of the Ostrogradski bootstrap. See Appendix B where a fourth order theory of the Strong interaction is defined that has color confinement and a linear r potential. The potential part of this theory was used by the Cornell group to calculate the Charmonium spectrum. (See Blaha (2017b) for details.)

An associated advantage of using PseudoQuantum Field Theory is that it provides for retarded propagators and an Arrow of Time.

---

[29] See Blaha (2017b) for the discussion of the PseudoQuantum field theory formalism for Higgs particles in our Extended Standard Model. See chapter 20 of Blaha (2017b), and earlier books, for a more detailed view than that presented here.

AND

4. Support for derivations of QUeST and UTMOST from one dimension BQUeST and BMOST.

We provide the author's paper, S. Blaha, Il Nuovo Cimento **49A**, 35 (1979) in the following pages for the reader's convenience (reproduced with the kind permission of Il Nuovo Cimento. Other papers from the 1970's are listed therein. Blaha (2007b) and (2011c) describe PseudoQuantum Field theory.

Blaha (2016f) describes PseudoQuantum Mechanics (*CQ Mechanics*), a related formalism, and describe applications (including the Schrödinger equation, Fokker-Planck equation, Boltzmann equation, and the Vlasov equation) that smoothly provide *continuous* transitions between Quantum Theory and Classical Theory.

IL NUOVO CIMENTO VOL. 49 A, N. 1 1 Gennaio 1979

The Local Definition of Asymptotic Particle States (*).

S. BLAHA

Physics Department, Williams College - Williamstown, Ma. 01267

(ricevuto il 28 Luglio 1978)

Summary. — A generalization of quantum field theory is described which has a unique particle interpretation even in space-times where no global timelike co-ordinate exists. The formulation is described in detail for the case of scalar bosons and spin-one-half fermions in flat space-time. We show that it is possible to construct a model in our approach which is physically equivalent to any given model in the usual formulation. In addition, a new class of models can be constructed which are not possible in the usual formulation. This class includes quantum action-at-a-distance models which can be used to develop models with higher-derivative field equations which are unitary. Our formulation allows some latitude in the choice of boundary conditions, so that one can opt for a continuum of possible Green's functions ranging from Feynman propagators to principal-value propagators (half advanced-half retarded).

1. – Introduction.

Our experience in flat space-time has fostered the opinion that a given action leads to a unique quantum field theory upon implementation of the canonical quantization procedure. This is apparently not true in general. A given action corresponds to an infinity of physically inequivalent quantum field theories in nonstatic space-times where no timelike Killing vector exists [1,2]. The origin of this plurality of quantum theories can be seen in free field the-

---

(*) Supported in part by grants from the National Science Foundation, and Research Corporation.

(1) S. A. FULLING: *Phys. Rev. D*, **7**, 2850 (1973); C. SOMMERFIELD: *Ann. Phys.*, **84**, 285 (1974).

(2) B. DEWITT: *Phys. Rep.*, **19**, 295 (1975).

ories (cf. FULLING (¹)). The usual quantization procedure is based on a definition of positive frequency which selects an acceptable complete orthonormal set of field equation solutions to use in field quantization. In nonstatic space-times no unique criterion exists for defining positive frequency. As a result, there is no restriction on the choice of complete orthonormal set of field equation solutions used to Fourier-expand fields. Having different choices leads to unitarily (and physically) inequivalent representations of the field algebra. The set of physical particle states in one quantization is generally not unitarily related to the set of physical states in another quantization (¹).

The absence of a criterion to select the « correct » quantum-field theory in the usual formulation has led us to consider a generalization of quantum field theory. In this generalization we introduce extra degrees of freedom in such a way that quantizations based on differing definitions of positive frequency are unitarily equivalent. Thus for a given action there is one resulting quantum field theory up to unitary equivalence.

In particular the physical particle states of different quantizations are related by a unitary transformation. Since the particle number operator is invariant under this transformation, a N-particle state in one quantization is a superposition of N-particle states in any other quantization. This is made possible by a local definition of particle states in the Fourier-transformed space (momentum space in the case of flat space-time).

It is important to note that the plethora of inequivalent quantizations in the usual formulation is faced by *one* observer. It is not a question of quantizations in different co-ordinate systems corresponding to different observers. The differences in the quantizations of two relatively accelerating observers, for example, are physically real and, in fact, also exist within the framework of our formulation. Relatively accelerating observers will, in general, « see » different numbers of particles.

Sections **2** and **3** contain our formulation of a free-scalar-boson field theory and a free spin-one-half fermion field theory in flat space-time. Significant differences exist between our formulation and the usual formulation. However, models exist in our formulation which make predictions which are identical to those of conventional field theory models, *e.g.*, quantum electrodynamics. Models also exist in our formulation which are completely outside the framework of the usual formulation. For example, a choice of boundary conditions is possible in our formulation which allows for virtual particles to propagate via non-Feynman propagators. In general, our particle propagator has the form

$$G = \sin^2\theta G_F + \cos^2\theta\, C G_F^* C^{-1}\,, \tag{1}$$

where θ is an arbitrary angle, G_F the usual Feynman propagator with G_F^* its complex conjugate, and where C is the relevant charge conjugation matrix. G is a Feynman propagator if $\theta = \pi/2$.

If $\theta = \pi/4$ then G is a principal-value propagator (half advanced-half retarded). This type of Green's function has appeared in classical action-at-a-distance theories. Our formulation thus encompases quantum action-at-a-distance theories. The use of principal-value propagators allows a substantial enlargement of the class of unitary, renormalizable field theory models. For example, models with higher-derivative field equations cannot simultaneously satisfy the requirements of positive probabilities and unitarity, if Feynman propagators are used. But if principal-value propagators are used, both requirements can be consistently satisfied (3). This has allowed us to previously construct a unitary, higher-derivative non-Abelian model of the strong interactions with a manifest linear potential and quark confinement (4). Of course, the use of principal-value propagators leads to a different type of analytic structure for amplitudes. We take the view that analyticity is an experimental question rather than a fundamental requirement on field theory (5). It is amusing to note that confinement of color in this model serves to sharply dampen if not eliminate the potential nonanalyticity.

We will discuss our formulation of non-Abelian field theories in detail in a subsequent paper (6).

2. – Boson quantization.

In flat space-time a timelike co-ordinate exists and as a result the Hamiltonian occupies a privileged position in defining positive frequency. In a nonstatic space-time, with no global timelike co-ordinate, no corresponding operator exists and the definition of « positive frequency » appears to be arbitrary. Consider a free scalar-field theory in such a situation. The field equation has an infinite number of possible complete orthonormal sets of solutions which span the space of solutions. Consider two possible sets: $\{\chi_\alpha, \chi_\alpha^*\}$ and $\{\psi_\beta, \psi_\beta^*\}$, where the χ_α are positive frequency with respect to one definition of positive frequency, and ψ_β are positive frequency with respect to a different definition. Then mode expansions of the scalar field

$$\varphi(x) = \sum_\alpha [\chi_\alpha(x) A_\alpha + \chi_\alpha^*(x) A_\alpha^\dagger], \tag{2}$$

$$\varphi(x) = \sum_\beta [\psi_\beta(x) b_\beta + \psi_\beta^*(x) b_\beta^\dagger], \tag{3}$$

---

(3) S. Blaha: *Phys. Rev. D*, **10**, 4268 (1974).
(4) S. Blaha: *Phys. Rev. D*, **11**, 2921 (1975).
(5) R. E. Cutkowsky, P. V. Landshoff, D. I. Olive and J. C. Polkinghorne: *Nucl. Phys.*, **12** B, 281 (1969); T. D. Lee: in *Quanta-Essays in Theoretical Physics Dedicated to Gregor Wentzel*, edited by P. G. O. Freund, C. J. Goebel and Y. Nambu (Chicago, Ill., 1970); H. Rechenberg and E. C. G. Sudarsham: *Nuovo Cimento*, **14** A, 299 (1973).
(6) S. Blaha: *Nuovo Cimento*, **49** A, 58 (1978).

can be inverted to relate the Fourier coefficient operators

$$A_\alpha = \sum_\beta [C_{\alpha\beta} b_\beta + \bar{C}_{\alpha\beta} b_\beta^\dagger], \tag{4}$$

where $C_{\alpha\beta}$ and $\bar{C}_{\alpha\beta}$ are c-number functions of x and β. Equation (4) shows that A_α is related to b_β and $b_\beta^\dagger$ through a local Bogoliubov transformation. As a result, the quantizations are, in general, not unitarily equivalent, have different vacua, and different particle interpretations [1,2]. The basis of this difficulty is the noncommutativity of Fourier coefficient operators and their Hermitian conjugates

$$[b_\beta, b_{\beta'}^\dagger] = \delta_{\beta\beta'} . \tag{5}$$

We shall propose a generalization of quantum field theory in which (in the free-field case) the Fourier coefficient operators and their Hermitian conjugates commute. In order to maintain the quantum character of the theory a supplementary field and the corresponding Fourier coefficient operators will be introduced. We shall confine our discussion to flat space-time in this section, and in sect. **3** which deals with free spin-one-half fermion quantization. In sect. **4** we discuss the particle interpretation of the formulation in nonstatic space-time.

Let us provisionally introduce the Lagrangian

$$\mathscr{L} = \partial_\mu \varphi_1 \partial^\mu \varphi_2 - \tfrac{1}{2} \partial_\mu \varphi_1 \partial^\mu \varphi_1 - m^2 \varphi_1 \varphi_2 + \tfrac{1}{2} m^2 \varphi_1^2 . \tag{6}$$

Following canonical procedures, we obtain the field equations

$$(\Box + m^2)\varphi_i = 0 , \tag{7}$$

for $i = 1, 2$ and the canonical momenta

$$\pi_1 = \dot{\varphi}_2 - \dot{\varphi}_1 \tag{8}$$

and

$$\pi_2 = \dot{\varphi}_1 , \tag{9}$$

which are taken to satisfy the canonical equal-time commutation relations

$$[\varphi_i(x), \pi_j(y)] = i\delta_{ij}\delta^3(\boldsymbol{x} - \boldsymbol{y}) , \tag{10}$$

$$[\varphi_i(x), \varphi_j(y)] = [\pi_i(x), \pi_j(y)] = 0 , \tag{11}$$

for $i, j = 1, 2$. Equations (9) and (10) imply

$$[\varphi_1(x), \dot{\varphi}_1(y)] = 0\,, \tag{12}$$

$$[\varphi_1(x), \dot{\varphi}_2(y)] = i\delta^3(\boldsymbol{x} - \boldsymbol{y})\,, \tag{13}$$

$$[\varphi_2(x), \dot{\varphi}_2(y)] = i\delta^3(\boldsymbol{x} - \boldsymbol{y})\,, \tag{14}$$

at equal times. The most general form for the mode expansion of the fields is

$$\varphi_1(x) = \int \mathrm{d}^3k\,[(C_{11}A_{1k} + C_{12}A_{2k})f_k(x) + (\bar{C}_{11}A^\dagger_{1k} + \bar{C}_{12}A^\dagger_{2k})f^*_k(x)]\,, \tag{15}$$

$$\varphi_2(x) = \int \mathrm{d}^3k\,[(C_{21}A_{1k} + C_{22}A_{2k})f_k(x) + (\bar{C}_{21}A^\dagger_{1k} + \bar{C}_{22}A^\dagger_{2k})f^*_k(x)]\,, \tag{16}$$

where $(2\pi)^{\frac{3}{2}}(2\omega_k)^{\frac{1}{2}}f_k(x) = \exp[-ik\cdot x]$ and where C_{ij} and $\bar{C}_{ij}$ are a set of constants. In view of the afore-mentioned difficulties stemming from the non-commutativity of a Fourier-coefficient operator and its Hermitian conjugate, we are led to impose the commutation relations

$$[A_{ik}, A_{jk'}] = [A^\dagger_{ik}, A^\dagger_{jk'}] = 0 \tag{17}$$

and

$$[A_{ik}, A^\dagger_{jk'}] = (1 - \delta_{ij})\,\delta^3(\boldsymbol{k} - \boldsymbol{k}')\,, \tag{18}$$

for $i, j = 1, 2$. We define two vacua (which are in fact related) $|0\rangle_1$ and $|0\rangle_2$ by

$$A_{1k}|0\rangle_2 = A^\dagger_{1k}|0\rangle_2 = 0\,, \tag{19}$$

$$A_{2k}|0\rangle_1 = A^\dagger_{2k}|0\rangle_1 = 0 \tag{20}$$

with

$$A_{2k}|0\rangle_2 \neq 0\,, \qquad A^\dagger_{2k}|0\rangle_2 \neq 0\,, \tag{21}$$

and

$$A_{1k}|0\rangle_1 \neq 0\,, \qquad A^\dagger_{1k}|0\rangle_1 \neq 0\,, \tag{22}$$

for all k. These definitions are motivated by the need for vacua which would be invariant under Bogoliubov transformations—a necessary requirement if the difficulties of particle interpretation caused by relations of the form of eq. (4) are to be avoided. Let us define the local Bogoliubov transformation

$$\begin{aligned} A_{ik}(\lambda_1, \lambda_2) &= B_{\lambda_1\lambda_2}A_{ik}B^{-1}_{\lambda_1\lambda_2} = \\ &= \exp[i\lambda_1]\cosh\lambda_2 A_{ik} + \exp[-i\lambda_1]\sinh\lambda_2 A^\dagger_{ik}\,, \end{aligned} \tag{23}$$

where λ_1 and λ_2 are functions of the momentum k. The operator B has the form

$$B_{\lambda_1\lambda_2} = \exp\left[2i\int d^3k\lambda_1(k)\Gamma_{3k}\right]\exp\left[2i\int d^3k\lambda_2(k)\Gamma_{2k}\right], \tag{24}$$

where

$$\Gamma_{3k} = (A^\dagger_{2k}A_{1k} + A_{2k}A^\dagger_{1k})/2, \tag{25}$$

$$\Gamma_{2k} = i(A^\dagger_{2k}A^\dagger_{1k} - A_{2k}A_{1k})/2. \tag{26}$$

If we also define

$$\Gamma_{1k} = -(A^\dagger_{2k}A^\dagger_{1k} + A_{2k}A_{1k})/2, \tag{27}$$

then these operators satisfy the commutation relations of a $SU_{1,1}$ algebra:

$$[\Gamma_{1k}, \Gamma_{2k'}] = -i\delta_{kk'}\Gamma_{3k}, \quad [\Gamma_{2k}, \Gamma_{3k'}] = i\delta_{kk'}\Gamma_{1k}, \quad [\Gamma_{3k}, \Gamma_{1k'}] = i\delta_{kk'}\Gamma_{2k}. \tag{28}$$

Thus the group of local Bogoliubov transformations is an infinite tensor product of $SU_{1,1}$ groups. It should be noted that $|0\rangle_2$ and $|0\rangle_1$ are invariant under this group. The equations of motion and equal-time commutation relations are also invariant under this group. These properties will enable us to show the uniqueness of the particle interpretation of our formulation in nonstatic space-time in sect. 4. The Casimir operator for the k-th $SU_{1,1}$ algebra,

$$\Gamma^2_k = \Gamma^2_{3k} - \Gamma^2_{1k} - \Gamma^2_{2k}, \tag{29}$$

$$\Gamma^2_k = N_k(N_k + 2), \tag{30}$$

allows us to identify the particle number operator (cf. sect. 4 below for its derivation)

$$N = \int d^3k N_k = \int d^3k (A^\dagger_{2k}A_{1k} - A_{2k}A^\dagger_{1k}), \tag{31}$$

which is left invariant by the Bogoliubov transformations.

If we compare our formulation to the usual one at this stage, we see that the enlargement of the field algebra has allowed us to define a group of local Bogoliubov transformations which is unitary and leaves the vacuum invariant —two properties not possible in the usual approach.

We now define inner products in our formalism. The structure of the commutation relations eq. (17) and (18) together with the nature of the defined vacua suggest that kets can be taken to have the form

$$|\alpha\rangle = A^\dagger_{2k_1}A^\dagger_{2k_2}\dots A_{2q_1}A_{2q_2}\dots|0\rangle_2 \tag{32}$$

and that bras should have the form

$$\langle\alpha| = {}_1\langle 0|A_{1k_1}A_{1k_2}\dots A^\dagger_{1q_1}A^\dagger_{1q_2}\dots. \tag{33}$$

(We could have chosen to construct kets using $|0\rangle_1$ and bras using ${}_2\langle 0|$ with no change in consequences.) The form of the commutation relations (which are used to reduce the inner products to a multiple of ${}_1\langle 0|0\rangle_2$) imply that the dual of the ket space is not its Hermitian conjugate. In our case the algebra reduces inner products to ${}_1\langle 0|0\rangle_2$ which we define to be unity.

We can relate the dual of a ket to its Hermitian conjugate through the introduction of a Dirac metric operator ([8]). We define the operator, γ, by

$$\gamma^{-1}A_{1k}\gamma = A_{2k}\,, \qquad \gamma^{-1}A_{2k}\gamma = A_{1k}\,, \tag{34}$$

$$\gamma|0\rangle_1 = |0\rangle_2\,. \tag{35}$$

We find

$$\gamma = \exp\left[-\frac{i\pi}{2}\int \mathrm{d}^3k\,[A_{2k}^\dagger A_{2k} + A_{1k}^\dagger A_{1k} - A_{2k}^\dagger A_{1k} - A_{2k}A_{1k}^\dagger]\right], \tag{36}$$

which implies that γ satisfies the necessary conditions for a metric operator, $\gamma = \gamma^\dagger = \gamma^{-1}$. The norm of a state $|\alpha\rangle$ can thus be defined by

$$(|\alpha\rangle)^\dagger\gamma|\alpha\rangle \tag{37a}$$

and inner products will generally have the form

$$(|\beta\rangle)^\dagger\gamma|\alpha\rangle\,. \tag{37b}$$

The adjoint operator is defined by

$$A^* = \gamma^{-1}A^\dagger\gamma\,. \tag{38}$$

Physical observables must be self-adjoint, $A^* = A$. Self-adjoint operators play the same role as Hermitian operators do in the usual formulation. In particular the Hamiltonian must be self-adjoint, if we are to have conservation of norm. Because φ_2 satisfies a Jordan-Pauli commutation relation, we shall introduce interactions in our model using only $\varphi_2(x)$. As a result φ_2 must also be self-adjoint if the Hamiltonian is to be self-adjoint.

([7]) Earlier two field formalisms have been considered by G. MIE: *Ann. Phys. Lpz.*, **37**, 511 (1912); P. A. M. DIRAC: *Comm. Dublin Inst. Advanced Studies*, **180** A, 1 (1942); W. PAULI: *Rev. Mod. Phys.*, **15**, 175 (1943); M. FROISSART: *Suppl. Nuovo Cimento*, **14**, 197 (1959); T. D. LEE and G. C. WICK: *Phys. Rev. D*, **2**, 1033 (1970). Our motivation and formulation differ substantially from them. Ref. ([3,4]) above do describe models which can be directly incorporated within the framework of our formulation.

([8]) W. PAULI: *Rev. Mod. Phys.*, **15**, 175 (1943).

The energy-momentum tensor is defined by

$$T^{\mu\nu} = -g^{\mu\nu}\mathscr{L} + \frac{\delta\mathscr{L}}{\delta\partial_\mu\varphi_1}\partial^\nu\varphi_1 + \frac{\delta\mathscr{L}}{\delta\partial_\mu\varphi_2}\partial^\nu\varphi_2 \tag{39}$$

with the Hamiltonian given by the 0-0 component. It is easy to verify that the requirements of Poincaré invariance, and the Schwinger commutation relations for $T^{\mu\nu}$ are met.

The equal-time commutation relations, and the self-adjointness of H and φ_2 place six constraints on the constants C_{ij} and $\bar{C}_{ij}$ in eqs. (15) and (16). After some algebra we find that we are able to express the field operators in the form

$$\varphi_1(x) = \int d^3k \left[\left(\frac{\cos(\theta_1-\theta_2)}{\sin\theta_1}A_{1k} + \frac{\sin(\theta_1-\theta_2)}{\sin\theta_1}A_{2k}\right)f_k(x) + \left(\frac{\cos(\theta_1-\theta_2)}{\cos\theta_1}A^\dagger_{1k} - \frac{\sin(\theta_1-\theta_2)}{\cos\theta_1}A^\dagger_{2k}\right)f^*_k(x)\right], \tag{40}$$

$$\varphi_2(x) = \int d^3k \left[(\cos\theta_2 A_{2k} + \sin\theta_2 A_{1k})f_k(x) + (\sin\theta_2 A^\dagger_{2k} - \cos\theta_2 A^\dagger_{1k})f^*_k(x)\right], \tag{41}$$

where θ_1 and θ_2 are arbitrary constants which fix the boundary conditions of the Green's functions. (They are *not* related to the Bogoliubov transformations defined above.) We also find

$$H = \int d^3k\,\omega_k(A^\dagger_{2k}A_{1k} + A_{2k}A^\dagger_{1k}) = 2\int d^3k\,\omega_k\Gamma_{3k} \tag{42}$$

in the free-field case independent of θ_1 and θ_2.

The theory is not invariant under Bogoliubov transformations due to their noncommutativity with H. This is consonant with the absence of any evidence in nature for such an invariance (and related constants of motion). The point of our formulation is to ensure that representations of the field algebra and dynamics, which are related to each other by Bogoliubov transformations, are unitarily equivalent. In the case of flat space-time the unitary equivalence is a moot point, since a unique generator of the dynamics, the Hamiltonian, is apparent. In nonstatic space-times, where no unique generator of the dynamical motion is determined, the unitary equivalence is necessary in order to have an unambiguous quantum field theory (given the action).

Different choices for the generator of the dynamics lead to representations which can be related by Bogoliubov transformations. These representations are unitarily equivalent in our formulation, but not equivalent in the usual formulation. We return to this issue in sect. 4.

The role of θ_1 and θ_2 is evident in the Green's functions. As usual we define the Green's functions as the vacuum expectation values of the time-ordered product of the field operators:

$$iG_{ij}(x-y) = {}_1\langle 0|T(\varphi_i(x)\varphi_j(y))|0\rangle_2 . \tag{43}$$

Equation (41) implies

$$G_{22}(x-y) = \sin^2\theta_2 G_F(x-y) - \cos^2\theta_2 G_F^*(x-y) , \tag{44}$$

where $G_F(x-y)$ is the usual Feynman propagator. G_{12} and G_{11} also depend on θ_1 and θ_2, but their precise expressions will not be of use in our presentation.

We shall now show that a model exists within our formulation which is physically equivalent to any conventional scalar quantum field theory with interaction $\mathscr{L}_I(\varphi)$. Our model Lagrangian is given by eq. (6) plus the interaction Lagrangian $\mathscr{L}_I(\varphi_2)$, where $\mathscr{L}_I(\varphi_2)$ is the same function of φ_2 as $\mathscr{L}_I(\varphi)$ is of φ. In order to have Feynman propagators, it is necessary to choose the boundary condition $\theta_2 = \pi/2$. We shall demonstrate that an asymptotic state exists in our formulation which corresponds to any asymptotic state of the usual formulation, and then show that S-matrix elements between corresponding states in the two models are equal in any order of perturbation theory.

The construction of asymptotic fields and states in our model is based on the renormalized quadratic part of the Lagrangian (eq. (6)). Therefore the previous development of this section can be used if appropriate subscripts « in » or « out » are appended to the operators. In particular, since $\theta_2 = \pi/2$, we have

$$\varphi_{2in}(x) = \int d^3k\,[f_k(x)A_{1kin} + f_k^*(x)A_{2kin}^\dagger] \tag{45a}$$

by eq. (41). We shall express the in-field operator of the usual formulation by

$$\varphi_{in}(x) = \int d^3k\,[f_k(x)A_{kin} + f_k^*(x)A_{kin}^\dagger] . \tag{45b}$$

Note that the form of φ_{2in} and φ_{in} is identical except for the subscripts « 1 » and « 2 » on the operators. In addition, the commutation relations of the field operators and the Fourier-coefficient operators are also identical except for numerical subscripts. Furthermore, the application of the field operators to the vacua is also identical in effect (except for subscripts)

$$\begin{cases} \varphi_{in}(x)|0\rangle = \varphi_{in}^{(-)}(x)|0\rangle , & \varphi_{2in}(x)|0\rangle_2 = \varphi_{2in}^{(-)}(x)|0\rangle_2 , \\ \langle 0|\varphi_{in}(x) = \langle 0|\varphi_{in}^{(+)}(x) , & {}_1\langle 0|\varphi_{2in}(x) = {}_1\langle 0|\varphi_{2in}^{(+)}(x) , \end{cases} \tag{45c}$$

where the superscript « + » labels positive-frequency parts of the field operator and « − » labels negative-frequency parts. This close parallel in properties between our model and the model of the usual formulation implies the identity

$$\langle 0|\mathscr{P}(\varphi_{in})|0\rangle = {}_1\langle 0|\mathscr{P}(\varphi_{2in})|0\rangle_2 , \tag{45d}$$

where $\mathscr{P}(\varphi_{1\text{in}})$ is any polynomial in the field $\varphi_{1\text{in}}$. Later we shall use this identity to demonstrate the equality of the S-matrices in our model and the given model of the usual formulation. (Note that a straightforward application of eq. (45*d*) implies that the propagator G_{22} in our formulation equals the time-ordered propagator of the usual formulation.)

We now state the rule associating asymptotic particle states in our formulation with those of the usual formulation: given an in or out ket of the usual formulation, the corresponding ket in our formulation is obtained by appending the subscript « 2 » to every Fourier-coefficient operator (and to the vacuum) (*e.g.* $A^\dagger_{k\text{in}}|0\rangle \Leftrightarrow A^\dagger_{2k\text{in}}|0\rangle_2$). Given an in or out bra of the usual formulation, the corresponding bra in our formulation is obtained by appending « 1 » to each Fourier-coefficient operator (and to the vacuum) (*e.g.* $\langle 0|A_{k\text{in}} \Leftrightarrow {}_1\langle 0|A_{1k\text{in}}$). It is easily seen that energy-momentum eigenstates in the usual formulation correspond to energy-momentum eigenstates in our formulation. Thus we have identified the set of physical states in our model and find a detailed correspondence to those of the usual formulation.

The development of the perturbation theory of our model is completely analogous to the usual development. The S-matrix relates in and out fields: $\varphi_{2\text{in}}(x) = S\varphi_{2\text{out}}(x)S^{-1}$. LSZ reduction formulae are derived in the same manner as in the usual formulation. We find the reduction formula for a particle from an in-state and from an out-state to be, respectively,

$$
(46)\quad
\begin{cases}
\langle \beta\,\text{out}|\alpha p\,\text{in}\rangle = \\
\qquad = \langle \beta - p\,\text{out}|\alpha\,\text{in}\rangle + \dfrac{i}{\sqrt{Z}}\displaystyle\int d^4x\, f_p(x)(\overrightarrow{\Box + m^2})\langle\beta\,\text{out}|\varphi_2(x)|\alpha\,\text{in}\rangle\,, \\[2ex]
\langle \beta k\,\text{out}|\alpha\,\text{in}\rangle = \\
\qquad = \langle \beta\,\text{out}|\alpha - k\,\text{in}\rangle + \dfrac{i}{\sqrt{Z}}\displaystyle\int d^4x\, f_k^*(x)(\overrightarrow{\Box + m^2})\langle\beta\,\text{out}|\varphi_2(x)|\alpha\,\text{in}\rangle\,,
\end{cases}
$$

where $\varphi_2(x)$ is the interacting field and where we use the notation of ref. [9]. The reduction of several particles leads to expressions which are identical to corresponding expressions of the usual model if the subscript « 2 » is appended to $\varphi(x)$.

Just as in the conventional model, we can formally develop a perturbation theory based on the U-matrix. The U-matrix relates the interacting and asymptotic field operator

$$
(47a)\qquad \varphi_2(\boldsymbol{x}, t) = U^{-1}(t)\varphi_{2\text{in}}(\boldsymbol{x}, t)U(t)
$$

([9]) We follow the conventions and notation of J. D. BJORKEN and S. D. DRELL: *Relativistic Quantum Fields* (New York, N. Y., 1965).

and is easily shown to satisfy the differential equation

$$(47b)\qquad i\frac{\partial U}{\partial t} = -\left[\int \mathrm{d}^3x\, \mathcal{L}_1(\varphi_{2in})\right] U\,.$$

Defining $U(t, t') = U(t)U^{-1}(t')$ and solving eq. (47b) gives

$$(48)\qquad U(t, t') = T\exp\left[i\int_{t'}^{t} \mathrm{d}^4x\, \mathcal{L}_1(\varphi_{2in})\right].$$

The LSZ procedure defined above reduces the calculation of S-matrix elements to the evaluation of time-ordered products of the interacting fields, ${}_1\langle 0|T(\varphi_2(x_1)\varphi_2(x_2)\dots\varphi_2(x_N))|0\rangle_2$. The U-matrix can then be used to reduce this quantity to the ratio of matrix elements involving only in-fields:

$$(49)\qquad \frac{{}_1\langle 0|T(\varphi_{2in}(x_1)\dots\varphi_{2in}(x_N)\exp[i\int \mathrm{d}^4x\, \mathcal{L}_1(\varphi_{2in})])|0\rangle_2}{{}_1\langle 0|T(\exp[i\int \mathrm{d}^4x\, \mathcal{L}_1(\varphi_{2in})])|0\rangle_2}\,.$$

Expanding to any order in the interaction in eq. (49) gives matrix elements of polynomials in φ_{2in} which are equal—term by term—to corresponding matrix elements of the perturbation theory of the model of the conventional formulation by eq. (45d). Thus S-matrix elements between corresponding states are identically equal in the conventional model and our corresponding model.

It should be noted that only a subset of the possible asymptotic states in our model are identified as physical particle states which correspond to states in the usual model. The operator A_{2kin} can also be used to create in-kets (and $A^{\dagger}_{1kout}$ to create out bras), but the S-matrix elements between physical kets and any ket (or bra) in which these operators appear is zero. (This follows from the fact that $[\mathcal{L}_1(\varphi_{2in}), A_{2kin}] = [\mathcal{L}_1(\varphi_{2in}), A^{\dagger}_{1kin}] = 0$ and ${}_1\langle 0|A_{2kin} = 0 = A^{\dagger}_{1kout}|0\rangle_2$.) Thus the S-matrix is block diagonal in our model. The part of it corresponding to the physical state sector is identical to the S-matrix of the given model of the conventional formulation.

The expression for the vacuum expectation value from which S-matrix elements may be calculated, eq. (49), can be used to show the unitary equivalence of representations which are related by a Bogoliubov transformation. Suppose that we had not used the representation of eq. (45a), but instead the Bogoliubov-transformed representation

$$(50)\qquad \varphi^{B}_{2in}(x) = \int \mathrm{d}^3k\, [f_k(x)(A_{1kin}\cosh\lambda + A^{\dagger}_{1kin}\sinh\lambda) + f^{*}_k(x)(A^{\dagger}_{2kin}\cosh\lambda + A_{2kin}\sinh\lambda)] \equiv B_{0\lambda}\varphi_{2in}(x)B^{-1}_{0\lambda}\,.$$

The canonical nature of the transformation guarantees that the canonical commutation relations will be maintained. If we follow the development of the perturbation theory given by eqs. (45)-(49) with φ^{B}_{2in} replacing φ_{2in} and $\varphi^{B}_{2} = U^{-1}(t)\varphi^{B}_{2in}U(t)$ replacing φ_2, then we find that S-matrix elements are calculated from vacuum expectation values involving only φ^{B}_{2in} fields:

$$(51)\qquad \frac{{}_1\langle 0|T(\varphi^{B}_{2in}(x_1)\dots\varphi^{B}_{2in}(x_N)\exp[i\int d^4x\,\mathscr{L}_1(\varphi^{B}_{2in})])|0\rangle_2}{{}_1\langle 0|T(\exp[i\int d^4x\,\mathscr{L}_1(\varphi^{B}_{2in})])|0\rangle_2}\,.$$

Since $B_{0\lambda}^{-1}|0\rangle_2 = |0\rangle_2$ and ${}_1\langle 0|B_{0\lambda} = {}_1\langle 0|$ we find that eq. (51) is equal to eq. (49). Thus the unitary equivalence of representations of the quantum field theory differing by a Bogoliubov transformation is demonstrated. (One can formally define Bogoliubov transformations for interacting fields $B^{int} = UBU^{-1}$, but B^{int} is not unitary due to the well-known difficulties of the U-matrix in the conventional formulation which are also present in our formulation. We circumvent this problem by working with the definition of the S-matrix in terms of vacuum expectation values of asymptotic in-fields, where the unitary equivalence under Bogoliubov transformation can be unambiguously shown to hold.)

A comparison of our formulation with the usual formulation shows a certain similarity of form at the Lagrangian level if our Lagrangian is put in the form

$$(52)\qquad \mathscr{L} = -\frac{1}{2}\partial_\mu(\varphi_1-\varphi_2)\partial^\mu(\varphi_1-\varphi_2) + \partial_\mu\varphi_2\partial^\mu\varphi_2 + \\ + \mathscr{L}_1(\varphi_2) + \frac{m^2}{2}(\varphi_1-\varphi_2)^2 - \frac{m^2}{2}\varphi_2^2\,.$$

In the usual approach $\varphi_3 = \varphi_1 - \varphi_2$ is an ignorable field and it would not have been surprising that we found equal S-matrix elements above. However, our formulation differs from the usual formulation in two respects—first, the field operators are both expanded in type «1» and «2» Fourier coefficient operators and, more importantly, the vacuum is defined in a way which correlates the φ_3 and φ_2 sectors. In the $\theta_2 = \pi/2$ case the first difference can be eliminated by a relabeling of Fourier-coefficient operators. However, for other values of θ_2 both differences are present and lead to a very different theory from the usual formulation. While it is clear that the correlation between the φ_3 and φ_2 sectors can be implemented in free field theory, one might ask if this remains true in the interacting case. Certainly the correlation can be implemented in the asymptotic fields and states, since that is free field theory. One can also *formally* implement the correlation for interacting fields through eq. (47a). But the implementation of the correlation in the interacting case is actually based on the reduction of the S-matrix element to the vacuum expectation value of products of asymptotic fields. Since the correlation can be maintained for the asymptotic fields and states, the physical quantities of the models, S-matrix elements, embody the correlation. (The value of the correlation we introduce is twofold: first, it is necessary in order to obtain the unitary equivalence of Bogoliubov rotated representations, and secondly, it widens the range of allowed flat–space-time quantum field theories to include those with principal-value propagators [8].)

We conclude this section with a brief discussion of our formulation of the charged-scalar-particle case. In the usual approach, the free-charged-scalar-particle Lagrangian may be expressed in terms of complex fields, $\varphi(x)$ and $\varphi^*(x)$ or in terms of two real fields $\varphi_a(x)$ and $\varphi_b(x)$ with

$$\varphi(x) = [\varphi_a(x) + i\varphi_b(x)]/\sqrt{2}\,. \tag{53}$$

If we follow the same procedure as above for the real fields, double their number and use the Lagrangian form of eq. (6), we are led to the complex field expression of the Lagrangian:

$$\mathscr{L} = \partial_\mu\bar{\varphi}_2\partial^\mu\varphi_1 + \partial_\mu\varphi_2\partial^\mu\bar{\varphi}_1 - \partial_\mu\bar{\varphi}_1\partial^\mu\varphi_1 - m^2\bar{\varphi}_2\varphi_1 - m^2\varphi_2\bar{\varphi}_1 + m^2\bar{\varphi}_1\varphi_1\,. \tag{54}$$

where

$$\varphi_i(x) = [\varphi_{ia}(x) + i\varphi_{ib}(x)]/\sqrt{2} \tag{55}$$

and

$$\bar{\varphi}_i(x) = [\varphi_{ia}(x) - i\varphi_{ib}(x)]/\sqrt{2}\,. \tag{56}$$

We require φ_{ia} and φ_{ib} to embody the same boundary conditions, so that the expansion of φ_{ia} and φ_{ib} utilizes the same constants, c_{ij} and $\bar{c}_{ij}$. Consequently

$$\varphi_i(x) = \int d^3k[(c_{i1}A_{+1k} + c_{i2}A_{+2k})f_k + (\bar{c}_{i1}A^\dagger_{-1k} + \bar{c}_{i2}A^\dagger_{-2k})f^*_k]\,, \tag{57}$$

$$\bar{\varphi}_i(x) = \int d^3k[(c_{i1}A_{-1k} + c_{i2}A_{-2k})f_k + (\bar{c}_{i1}A^\dagger_{+1k} + \bar{c}_{i2}A^\dagger_{+2k})f^*_k]\,, \tag{58}$$

where φ_i and $\bar{\varphi}_i$ are related via the charge conjugation operator [9]. Following the quantization pattern discussed above, with only minor changes due to the presence of two types of Fourier coefficient operators: positive charge, A_{+ik}, and negative charge, A_{-ik}, leads eventually to the following Green's function:

$$G_{22}(x-y) = G_F(x-y)\sin^2\theta_2 + G^*_F(x-y)\cos^2\theta_2\,. \tag{59}$$

Note that it has the same form as eq. (1). (Lagrangian interaction terms are expressed solely in terms of φ_2 and $\bar{\varphi}_2$.) As a result we require $\gamma\bar{\varphi}_2\gamma^{-1} = \varphi_2^\dagger$, the Hermitian conjugate of φ_2, where γ is the metric operator so that

$$C\varphi_2C^{-1} = \gamma\varphi_2^\dagger\gamma^{-1}\,, \tag{60}$$

where C is the charge conjugation operator.

3. – Fermion quantization.

In this section we describe our formulation of spin-one-half fermion quantum field theory. Again we are motivated by the need for a unique particle interpretation in nonstatic space-time. The formulation for fermions has close similarities to boson quantization.

Two fields are needed to describe a spin-one-half particle. The Lagrangian is

$$\mathscr{L} = \tilde{\psi}_2\gamma^0(i\nabla - m)\psi_1 + \tilde{\psi}_1\gamma^0(i\nabla - m)\psi_2 - \tilde{\psi}_1\gamma^0(i\nabla - m)\psi_1 . \tag{61}$$

We follow the conventions and notation of ref. (9). The fields $\tilde{\psi}_i$ will be related to the transpose of the charge conjugate field via

$$\tilde{\psi}_i = \psi_i^{cT}\gamma^0 C^T \tag{62}$$

for $i = 1, 2$. (In the usual formulation $\psi^\dagger = \tilde{\psi}$ would hold.) The equations of motion are

$$(i\nabla - m)\psi_i = 0 , \qquad \tilde{\psi}_i(i\overleftarrow{\nabla} - m) = 0 , \tag{63}$$

for $i = 1, 2$. The momentum conjugate to ψ_1 is

$$\pi_1 = i(\tilde{\psi}_2 - \tilde{\psi}_1) \tag{64}$$

and the conjugate to ψ_2 is

$$\pi_2 = i\tilde{\psi}_1 . \tag{65}$$

The canonical equal-time anticommutation relations imply

$$\{\psi_{1\alpha}(x), \tilde{\psi}_{1\beta}(y)\} = 0 , \tag{66}$$

$$\{\psi_{1\alpha}(x), \tilde{\psi}_{2\beta}(y)\} = \delta_{\alpha\beta}\delta^3(\boldsymbol{x} - \boldsymbol{y}) \tag{67}$$

and

$$\{\psi_{2\alpha}(x), \tilde{\psi}_{2\beta}(y)\} = \delta_{\alpha\beta}\delta^3(\boldsymbol{x} - \boldsymbol{y}) . \tag{68}$$

The most general form for the mode of expansion of the fields is (9)

$$\psi_i = \sqrt{2m}\sum_s \int \mathrm{d}^3k\,[(c_{i1}b_{1ks} + c_{i2}b_{2ks})f_k(x)u_{ks} + (\tilde{c}_{i1}d^\dagger_{1ks} + \tilde{c}_{i2}d^\dagger_{2ks})f^*_k(x)v_{ks}] . \tag{69}$$

Just as in the charged scalar case, we develop our formulation in such a way that the even and odd charge conjugation combinations, $\psi_i \pm \psi_i^c$, implement the same boundary conditions. Therefore

$$(70)\qquad \tilde{\psi}_i = \sqrt{2m}\sum_s \int d^3k[(c_{i1}d_{1ks} + c_{i2}d_{2ks})f_k(x)v_{ks}^\dagger + (\tilde{c}_{i1}b_{1ks}^\dagger + \tilde{c}_{i2}b_{2ks}^\dagger)f_k^*(x)u_{ks}^\dagger].$$

The nonzero Fourier-coefficient anti-commutation relations are

$$(71)\qquad \{d_{iks}, d_{jk's'}^\dagger\} = \{b_{iks}, b_{jk's'}^\dagger\} = (1-\delta_{ij})\delta_{ss'}\delta^3(k-k')$$

for $i, j = 1, 2$. The definition of states and inner products mirror the boson case. The vacua are defined by

$$(72)\qquad b_{1ks}|0\rangle_2 = b_{1ks}^\dagger|0\rangle_2 = d_{1ks}|0\rangle_2 = d_{1ks}^\dagger|0\rangle_2 = 0$$

and

$$(73)\qquad b_{2ks}|0\rangle_1 = b_{2ks}^\dagger|0\rangle_1 = d_{2ks}|0\rangle_1 = d_{2ks}^\dagger|0\rangle_1 = 0$$

and are related by a metric operator η:

$$(74)\qquad \eta|0\rangle_1 = |0\rangle_2$$

which satisfies $\eta = \eta^\dagger = \eta^{-1}$. We conventionally choose to construct kets from $|0\rangle_2$ and define their dual as their Hermitian conjugate multiplied by the metric operator. Thus inner products have the form

$$(75)\qquad \langle\alpha|\beta\rangle = (|\alpha\rangle)^\dagger\eta|\beta\rangle.$$

Physical observables must be self-adjoint, $A = A^* = \eta^{-1}A^\dagger\eta$, in order to have real eigenvalues. The Hamiltonian must be self-adjoint in order to conserve the norm. In view of eq. (68) we only use ψ_2 and $\tilde{\psi}_2$ in interaction terms and therefore require $\tilde{\psi}_2 = \eta^{-1}\psi_2^\dagger\eta$, so that

$$(76)\qquad C\psi_2 C^{-1} = \eta^{-1}C\tilde{\psi}_2^T\eta,$$

which bears comparison with eq. (15.112) of ref. (9) and also eq. (60) above. The equal-time anticommutation relations, eq. (76), and the adjointness of H restrict the constants c_{ij} and $\tilde{c}_{ij}$ so that

$$(77)\qquad \psi_1 = \sqrt{2m}\sum_s\int d^3k[(\cos(\theta_1-\theta_2)b_{1ks} + \sin(\theta_1-\theta_2)b_{2ks})f_k(x)u_{ks}\sin\theta_1 + \\ + (\cos(\theta_1-\theta_2)d_{1ks}^\dagger - \sin(\theta_1-\theta_2)d_{2ks}^\dagger)f_k^*(x)v_{ks}/\cos\theta_1]$$

and

$$(78)\qquad \psi_2 = \sqrt{2m}\sum_s\int d^3k[(\sin\theta_2 b_{1ks} + \cos\theta_2 b_{2ks})f_k(x)u_{ks} + \\ + (\cos\theta_2 d_{1ks}^\dagger + \sin\theta_2 d_{2ks}^\dagger)f_k^*(x)v_{ks}].$$

The Hamiltonian is

$$H = \sum_s \int d^3k\,\omega_k (b^\dagger_{2ks} b_{1ks} - b_{2ks} b^\dagger_{1ks} + d^\dagger_{2ks} d_{1ks} - d_{2ks} d^\dagger_{1ks})\,. \tag{79}$$

In contrast to the usual formulation, we see that our Hamiltonian does not have an infinite vacuum energy with respect to $|0\rangle_2$. It is not positive definite, but we will be able to develop a unitary S-matrix theory in the space of positive-energy asymptotic states, if we choose $\theta_2 = \pi/2$. This is evident from an examination of the Green's function

$$S_{22}(x-y) = -i\,{}_1\langle 0|T(\psi_2(x)\,\overline{\psi}_2(y)\gamma_0)|0\rangle_2 = \tag{80}$$

$$= \sin^2\theta_2 S_F(x-y) + \cos^2\theta_2\, C\gamma^0 S^*_F (C\gamma^0)^{-1}\,, \tag{81}$$

which gives $S_{22} = S_F$, the usual Feynman propagator, if $\theta_2 = \pi/2$. As in the boson case, we introduce interactions only through type-2 operators, and use type-2 operators to LSZ reduce in and out particles. The result is a perturbation theory which, in the Fermion sector, only involves time-ordered products of ψ_2 and $\overline{\psi}_2$. Thus we can establish a model in our formulation which is equivalent, so far as S-matrix elements are concerned, to any given model of the usual formulation. In particular, our model electrodynamics has the Lagrangian

$$\mathscr{L} = \tfrac{1}{2}F^1_{\mu\nu}F^{2\mu\nu} + \tfrac{1}{4}F^1_{\mu\nu}F^{1\mu\nu} - \overline{\psi}_2\gamma^0(i\nabla - m)\psi_1 + \tag{82}$$
$$+ \overline{\psi}_1\gamma^0(i\nabla - m)\psi_2 - \overline{\psi}_1\gamma^0(i\nabla - m)\psi_1 - e_0\overline{\psi}_2\gamma^0\hat{A}_2\psi_2$$

with

$$F^i_{\mu\nu} = \partial_\nu A_{i\mu} - \partial_\mu A_{i\nu} \tag{83}$$

for $i = 1, 2$. While we will discuss this model more fully elsewhere [6] two things are worth noting. First the interaction is expressed solely in terms of fields of type 2—both for fermions and the photon. Following our quantization procedure leads to S-matrix expressions which are term-by-term equal to corresponding expressions in QED. Secondly, the model is gauge invariant. The gauge transformation is

$$\psi_2 \to \exp[i\Lambda]\psi_2\,, \tag{84}$$

$$\overline{\psi}_2 \to \exp[-i\Lambda]\overline{\psi}_2\,, \tag{85}$$

$$A_{2\mu} \to A_{2\mu} - \frac{1}{e}\partial_\mu\Lambda\,, \tag{86}$$

$$\psi_1 \to \psi_1 + (\exp[i\Lambda] - 1)\psi_2\,, \tag{87}$$

$$\overline{\psi}_1 \to \overline{\psi}_1 + (\exp[-i\Lambda] - 1)\overline{\psi}_2\,, \tag{88}$$

and its associated conserved current is

$$J_\mu = -i\frac{\delta\mathscr{L}}{\delta\bar{\partial}^\mu\psi_1}\psi_2 - i\frac{\delta\mathscr{L}}{\delta\bar{\partial}^\mu\psi_2}\psi_2\,, \tag{89}$$

$$J_\mu = \bar{\psi}_2\gamma^0\gamma_\mu\psi_2\,. \tag{90}$$

We close our discussion of fermions by considering the case $\theta_2 = \pi/4$ which, by eq. (81), gives the principal-value propagator

$$S_{22}(x-y) = \int\frac{d^4k}{(2\pi)^4}\exp[-ik\cdot(x-y)](k+m)\frac{P}{k^2-m^2}\,, \tag{91}$$

where

$$\frac{P}{k^2-m^2} = \frac{1}{2}\left(\frac{1}{k^2-m^2+i\varepsilon} + \frac{1}{k^2-m^2-i\varepsilon}\right). \tag{92}$$

Thus a quantum action-at-a-distance model of fermions can be constructed within the framework of our formulation.

4. – Particle interpretation.

In this section we shall show that the particle interpretation of our formulation of quantum field theory is well defined for the case of a free scalar particle in a nonstatic space-time where no global timelike co-ordinate exists. We assume an action of the form

$$S = \int d^4x\,[\varphi_2 D\varphi_1 - \tfrac{1}{2}\varphi_1 D\varphi_1] + (\text{surface terms})\,, \tag{93}$$

which under the variation of S gives the field equations

$$D\varphi_1 = D\varphi_2 = 0\,. \tag{94}$$

The self-adjointness of D implies

$$\int_V [f^* Dg - (Df)^* g]\,d^4x = \int_{\Sigma_v} f^*\overleftrightarrow{D}^\mu g\,d\Sigma_\mu\,, \tag{95}$$

where Σ_v is the surface bounding V, $d\Sigma_\mu$ is an outward directed surface element of Σ_v and D^μ is a two-edged vector differential operator. If Σ is a spacelike complete Cauchy hypersurface for the field equations (we assume they exist), then an inner product for complex solutions of the field equations can be defined by

$$(v_1, v_2) = i \int_\Sigma v_1^* \overleftrightarrow{D}^\mu v_2 \, d\Sigma_\mu . \tag{96}$$

We now choose an arbitrary complete orthonormal set of pairs of complex conjugate solutions of eq. (94), $\{V_\alpha, V_\alpha^*\}$, satisfying

$$(V_\alpha, V_{\alpha'}) = -(V_\alpha^*, V_{\alpha'}^*) = \delta_{\alpha\alpha'} , \tag{97}$$

$$(V_\alpha, V_{\alpha'}^*) = 0 , \tag{98}$$

and use them in the mode expansion of the field operators

$$\varphi_i = \sum_\alpha [(c_{i1} A_{1\alpha} + c_{i2} A_{2\alpha}) V_\alpha + (\bar{c}_{i1} A_{1\alpha}^\dagger + \bar{c}_{i2} A_{2\alpha}^\dagger) V_\alpha^*] , \tag{99}$$

where c_{ij} and $\bar{c}_{ij}$ are real c-numbers. The Fourier coefficient operators satisfy

$$[A_{i\alpha}, A_{j\alpha'}^\dagger] = (1 - \delta_{ij}) \delta_{\alpha\alpha'} \tag{100}$$

with all other commutators equal to zero. The commutativity of Fourier-coefficient operators and their Hermitian conjugate allows us to define the vacua $|0\rangle_1$ and $|0\rangle_2$ by

$$A_{1\alpha}|0\rangle_2 = A_{1\alpha}^\dagger|0\rangle_2 = A_{2\alpha}|0\rangle_1 = A_{2\alpha}^\dagger|0\rangle_1 = 0 . \tag{101}$$

We choose to construct states from $|0\rangle_2$. The one-particle ket corresponding to the Fourier transform variable α is

$$|\alpha\rangle = -(v_\alpha^*, \varphi_2)|0\rangle_2/\bar{c}_{22} \tag{102}$$

and the one-particle bra dual to it is

$$\langle\alpha| = (|\alpha\rangle)^\dagger \gamma , \tag{103}$$

where γ is a metric operator satisfying $\gamma = \gamma^{-1} = \gamma^\dagger$ and

$$\gamma^{-1} A_{i\alpha} \gamma = \varepsilon_{ij} A_{j\alpha} \tag{104}$$

with $\varepsilon_{11} = \varepsilon_{22} = 0$ and $\varepsilon_{12} = \varepsilon_{21} = 1$. The further development of this quantization proceeds along the lines of sect. **2**. In particular φ_2 is self-adjoint.

We now introduce a quantization of the particle described by S which parallels the above development in every detail except that a different complete orthonormal set of field equation solutions $\{W_\beta, W_\beta^*\}$ is used in the mode expansion of the fields

$$\varphi_i = \sum_\beta \left[(c_{i1} A_{1\beta} + c_{i2} A_{2\beta}) W_\beta + (\bar{c}_{i1} A^\dagger_{1\beta} + \bar{c}_{i2} A^\dagger_{2\beta}) W^*_\beta \right]. \tag{105}$$

The question arises: how are we to relate the two quantizations? In the usual formulation only one answer is apparent—the field operators are to be identified [1,2], since they are uniquely determined by the field equations and the canonical commutation relations. But in the present case, the field operators are not uniquely determined, so that the identification of the fields in eq. (105) and (99) is not required. The relation between the quantizations must obviously be well defined (in the sense that every operator and state in one quantization can be uniquely expressed in terms of operators and states of the other representations). More importantly, it must only relate operators whose properties are fixed by the field equation and the canonical commutation relations; and whose expectation values are uniquely specified by purely geometrical restrictions on their support and do not embody a definition of positive frequency. In our formalism, the operators which satisfy these requirements are linear combinations of

$$\varphi^{\text{II}}_{iv} = \sum_\alpha (A_{i\alpha} V_\alpha + A^\dagger_{i\alpha} V^*_\alpha), \tag{106}$$

$$\varphi^{\text{II}}_{iw} = \sum_\beta (A_{i\beta} W_\beta + A^\dagger_{i\beta} W^*_\beta), \tag{107}$$

for $i = 1, 2$ in the respective quantizations we can restrict the discussion to these quantities. In particular, the vacuum expectation value

$${}_1\langle 0|\varphi^{\text{II}}_1(x)\varphi^{\text{II}}_2(y)|0\rangle_2 = \tfrac{1}{2}\,{}_1\langle 0|[\varphi^{\text{II}}_1(x), \varphi^{\text{II}}_2(y)]|0\rangle_2, \tag{108}$$

$${}_1\langle 0|\varphi^{\text{II}}_1(x)\varphi^{\text{II}}_2(y)|0\rangle_2 = \frac{i}{2}\Delta(x-y), \tag{109}$$

where $\Delta(x-y)$, the commutator function, has the well-defined geometrical property that it vanishes at spacelike distances.

Identifying φ^{II}_{iw} with φ^{II}_{iv}, for $i = 1, 2$, leads to the relations

$$A_{i\beta} = (W_\beta, \varphi^{\text{II}}_i) = \sum_\alpha \left[(W_\beta, V_\alpha) A_{i\alpha} + (W_\beta, V^*_\alpha) A^\dagger_{i\alpha} \right], \tag{110}$$

for $i = 1, 2$ plus Hermitian-conjugate expressions. The form of the inner products on the right-hand side of eq. (110) is determined by requiring that the definition of positive frequency implicit in the separation of the orthonormal set $\{W_\beta, W^*_\beta\}$ into complex conjugate pairs of solutions can also be implemented by linear combinations of V_α and V^*_α. Specifically, we assume that a complete orthonormal set of pairs of complex conjugate functions, $\{V_\alpha, V^*_\alpha\}$, exists which satisfies

$$V_\alpha = c_1 \tilde{V}_\alpha + c_2 \tilde{V}^*_\alpha, \tag{111}$$

$$V^*_\alpha = c^*_1 \tilde{V}^*_\alpha + c^*_2 \tilde{V}_\alpha. \tag{112}$$

for all α, where c_1 and c_2 are c-number functions of α only with $|c_1| > |c_2|$, and where

$$(W_\beta, \tilde{V}^*_\alpha) = 0, \tag{113}$$

$$(W^*_\beta, \tilde{V}_\alpha) = 0, \tag{114}$$

for all β. The orthogonality conditions imply

$$c_1 = \exp[i\lambda_1] \cosh \lambda_2, \tag{115}$$

$$c_2 = \exp[i\lambda_1] \sinh \lambda_2. \tag{116}$$

where λ_1 and λ_2 are solely functions of α. The substitution of eqs. (111) and (112) in eq. (110) and use of eqs. (113) and (114) gives

$$A_{i\beta} = \sum_\alpha (W_\beta, \tilde{V}_\alpha)[\exp[i\lambda_1] \cosh \lambda_2 A_{i\alpha} + \exp[-i\lambda_1] \sinh \lambda_2 A^\dagger_{i\alpha}] \tag{117}$$

for $i = 1, 2$. Note that the bracketed term on the right-hand side of the equation has the same form as the Bogoliubov rotated Fourier-coefficient operator given in eq. (23). In the present case we can rewrite eq. (117) in the form

$$A_{i\beta} = \sum_\alpha (W_\beta, \tilde{V}_\alpha) B_{\lambda_1\lambda_2} A_{i\alpha} B^{-1}_{\lambda_1\lambda_2} \tag{118}$$

with

$$B_{\lambda_1\lambda_2} = \exp\left[2i \sum_\alpha \lambda_1(\alpha) \Gamma_{3\alpha}\right] \exp\left[2i \sum_\alpha \lambda_2(\alpha) \Gamma_{2\alpha}\right]. \tag{119}$$

where $\Gamma_{3\alpha}$ and $\Gamma_{2\alpha}$ are obtained from eqs. (25) and (26) by replacing the subscripts k with α.

The particle interpretations of the two quantizations will now be shown to be identical. First we note that the vacuum $|0\rangle_2$ of the « α » quantization is invariant under $B_{\lambda_1\lambda_2}$, so that it may be taken to be identical with the $|0\rangle_2$ vacumm of the « β » quantization. Next we note that the canonical commutation relations and the vacuum expectation value of any product of field operators are invariant under B:

$${}_1\langle 0|\varphi_{i_1}(x_1)\varphi_{i_2}(x_2) \dots |0\rangle_2 = {}_1\langle 0|B^{-1}\varphi_{i_1}(x_1)\varphi_{i_2}(x_2) \dots B|0\rangle_2 . \tag{120}$$

This implies that we could replace $A_{i\alpha}$ with $B_{\lambda_1\lambda_2}A_{i\alpha}B^{-1}_{\lambda_1\lambda_2}$ in the mode expansions, eq. (107), with no change in physical consequences. In particular, this applies to the definition of particle kets. Equation (102) becomes

$$|\alpha\rangle_1 = (\exp[-i\lambda_1]\cosh\lambda_2 A^{\dagger}_{2\alpha} + \exp[i\lambda_1]\sinh\lambda_2 A_{2\alpha})|0\rangle_2 . \tag{121}$$

Consequently, $A^{\dagger}_{2\beta}|0\rangle_2$ is a superposition of one-particle states in the « α » quantization. In general, the N-particle state in the « β » quantization is a superposition of N-particle states in the « α » quantization.

The invariance of particle number under Bogoliubov transformations is reflected in the relation between the particle number operator,

$$N = \sum_{\alpha}(A^{\dagger}_{2\alpha}A_{1\alpha} - A_{2\alpha}A^{\dagger}_{1\alpha}) , \tag{122}$$

which is invariant under Bogoliubov transformations, and related to the Casimir operator of the Bogoliubov group (cf. eqs. (29)-(31)). Our identification of N as the particle number operator is based, as most charge and number operators are, on an invariance of the action under a global change of phase of fields. In our case we note that the action of eq. (93) is invariant under the infinitesimal phase change

$$\varphi_1 \to \varphi_1 + i\varepsilon\varphi_1 , \qquad \varphi_2 \to \varphi_2 + i\varepsilon(\varphi_1 - \varphi_2) . \tag{123}$$

The corresponding conserved-number operator is given by

$$N = i\int_{\Sigma_\nu}\varphi_1 \overleftrightarrow{D}^{\mu}\varphi_2 \, d\Sigma_\mu . \tag{124}$$

Because φ_1 and φ_2 implicitly embody a definition of positive frequency, we are led to replace them with Hermitian operators:

$$N = i\int_{\Sigma_\nu}\varphi^{\rm H}_1 \overleftrightarrow{D}^{\mu}\varphi^{\rm H}_2 \, d\Sigma_\mu \tag{125}$$

with $\varphi^{\rm H}_i$ given in eq. (106). Equation (125) can be evaluated by using eq. (96) and (97) to give eq. (122). Thus our definition of number operator is physically motivated. It is also consistent with our expectations of a number operator.

We shall now summarize our picture of second quantization in curved space-time where no global timelike Killing vector is present. Consider a complete spacelike Cauchy hypersurface. At each point on the surface there is a local timelike direction. There is, in general, a class of operators, which will locally generate a displacement in the timelike direction, but which globally generate very different motions. Due to the absence of a global timelike Killing vector, no member of the class of potential generators of the dynamics is physically selected as the generator of the dynamics. One is free to choose any member as the generator of the dynamics locally. Each choice implies a different definition of positive and negative frequency when field operators are represented by Fourier expansions.

In the usual formulation of quantum field theory any choice of generator of the dynamics (and thus Fourier representation of field operators) is unitarily inequivalent to any other choice in general. As a result each choice gives a *different physical theory* and the second quantization of a theory is not unique. Practically, this means that 1) a one-particle state in one quantization is a many-particle state in any other quantization (particle number is ambiguous), 2) in general one can construct a one-particle state which is an eigenstate of a generator in one representation, but one cannot construct a one-particle state in another representation which is an eigenstate of the same generator (the space of states is different), and 3) (if interactions are introduced) the S-matrix differs from quantization to quantization. Obviously, there are only two acceptable alternatives in this situation; either some new principle selects one representation as the correct physical representation, or a modification of quantum field theory is necessary. In the absence of a new physical principle, we have formulated a modification of quantum field theory.

Our formulation allows one to quantize a field theory with any of the potential generators of the dynamics and yet to have a physically unique theory. Different quantizations can be related by Bogoliubov transformations and, in our formulation, are unitarily equivalent. Consequently, particle number is invariant—N-particle states in one representation are superpositions of N-particle states in any other representation; the set of states in one representation is unitarily equivalent to the set of states in any other representation: and the S-matrix is uniquely determined in the case of an interacting theory (the proof is analogous to that of the flat space-time case discussed in sect. **2**). Our formulation associates a unique physical theory with any given action. In a sense, it implements an equivalence principle in the space of solutions to the field equations—any complete orthonormal set of solutions to the field equations can be used in the Fourier expansion of field operators and a unique physical theory results (cf. eqs. (111)-(114)).

In conclusion, we note that the problem we have addressed relates to one observer and the ambiguities of conventional quantum field theory he must face. Different observers in relatively accelerating frames will not see the same number of particles in our formulation. Neither is particle creation near black holes precluded in our formulation.

* * *

I am grateful to the Aspen Center for Physics for its hospitality while part of this work was being done, and to M. A. B. BEG, S. MANDELSTAM and D. PARK for stimulating conversations.

● RIASSUNTO (*)

Si descrive una generalizzazione della teoria quantistica dei campi che ha un'unica interpretazione particellare — anche negli spazio-tempo in cui non esiste alcuna coordinata globale di tipo tempo. La formulazione è descritta in dettaglio per i casi di bosoni scalari e di fermioni con spin ½ nello spazio-tempo piatto. Si mostra che è possibile costruire un modello nel nostro approccio che è fisicamente equivalente a un qualsiasi modello nella solita formulazione. Inoltre, si può costruire una nuova classe di modelli che non sono possibili nella solita formulazione. Questa classe comprende modelli quantici di azione a distanza che possono essere usati per sviluppare modelli con equazioni di campo a derivata più alta che sono unitarie. La nostra formulazione permette ampia scelta delle condizioni limite, cosicché si può optare per un continuo di possibili funzioni di Green che vanno dai propagatori di Feynman a propagatori del valore principale (mezzo avanzato-mezzo ritardato).

(*) *Traduzione a cura della Redazione.*

Локальное определение асимптотических состояний частиц.

Резюме (*). — Описывается обобщение квантовой теории поля, которая имеет единую частичную интерпретацию — даже в пространстве и времени, где не существует глобальной времениподобной координаты. Подробно описывается формулировка для случаев скалярных бозонов и фермионов со спином половина в плоском пространстве-времени. Мы показываем, что имеется возможность сконструировать модель в нашем подходе, которая физически эквивалентна любой заданной модели в обычной формулировке. Кроме того, может быть сконструирован новый класс моделей, которые являются невозможными в обычной формулировке. Этот класс включает модели квантового действия на расстоянии, которые могут быть использованы для развития моделей с полевыми уравнениями с высшими производными, которые являются унитарными. Наша формулировка допускает некоторую свободу в выборе граничных условий, так что имеется возможность выбрать континуум возможных гриновских функций, от фейнмановских пропагаторов до пропагаторов главных значений (наполовину опережающая - наполовину запаздывающая).

(*) *Переведено редакцией.*

2. UST Derived from QUeST

BQUeST, QUeST, and UST form a complete theory of elementary particles and Gravitation. In chapter 1 we derived QUeST from one dimension BQUeST. In this chapter we show that QUeST implies UST: same internal symmetries, same fermion, vector boson, and Higgs particles. *This chapter shows that starting from BQUeST with one fermion with one internal dimension residing in an 8 dimension Megaverse one can generate QUeST and show it is consistent with UST.*

The probability of a consistent match of quaternion-based QUeST and logic-based UST would seem to be very low. The consistency suggests that we are on the right track for a complete theory of elementary particles and gravitation.

2.1 Internal Symmetries

QUeST has a 256 dimension space. This space furnishes a fundamental representation for U(128). This symmetry must be broken to obtain the internal symmetries and space-time of UST (and of the Standard Model within it). A pattern of symmetry breaking[30] that leads to UST and the Standard Model is:

1. U(128) broken to U(32)⊗U(32)⊗U(32)⊗U(32) giving four layers
2. Each U(32) broken to a 32 dimension U(16) giving "Normal" and "Dark" sectors
3. Each U(16) broken to a 16 dimension U(8) giving subdivisions of "Normal" and "Dark"
4. Each U(8) broken to SU(2)⊗U(1)⊗SU(3)⊗U(1)⊗U(4)⊗U(4)⊗(4 dimension space-time part)

Figure 2.1. Pattern of symmetry breaking of QUeST U(128).

This pattern results in a "tiling" of the 256 dimension QUeST array with 16 tiles composed of the groups:

$$SU(2)\otimes U(1)\otimes SU(3)\otimes U(1)\otimes(4 \text{ dimension space-time part}) \quad \text{or} \quad U(4)\otimes U(4) \qquad (2.1)$$

The space-time parts combine to give a four complex quaternion space-time from which are real four dimension space-time results. One might ask how our real space-time results. One simple answer is that it is the result of our real four dimension measuring tools. We can only "see" things of our four dimensions ultimately because of conservation of momentum. Things within our space-time are visible. Things outside our space-time are not. Things partly within and partly without would be thought to have anomalous features (such as anomalous mass), and would usually be discarded as

[30] Suggested by the derivation of QUeST in a two-step process: first using the one dimension fermion's four spinors (the fermion is in a 4-dimension external space) to boost 1-dimension BQUeST to a 4 × 4 array; then assuming each of the array components is a fermion using each's 4-spinors boost to obtain a 16 × 16 dimension QUeST array. See Blaha (2020i) for details.

experimental, unreproducible errors. In the case of the universe, distortions in its structure could be attributed to objects beyond the universe.

The symmetry breaking of Fig. 2.1 lead to the internal symmetries of QUeST:

$$[SU(2)\otimes U(1)\otimes SU(3)\otimes U(1)]^8\otimes U(4)^{16} \tag{2.2}$$

The corresponding layout of the QUeST dimensions appears in Fig. 2.2. There are four layers of internal symmetries in QUeST. Each layer has two SU(2)⊗U(1)⊗SU(3)⊗U(1) groups, two U(4) Generation groups, two U(4) Layer groups, and two U(1) Fermion groups (described later).[31] See Figs. 2.2 and 2.3.

[31] We have replaced the Dark U(2) group seen earlier here and in previous books with a pair of U(1) groups because we were better able to achieve the $4 \times 4 = 16$ dimensions pattern that seems present.

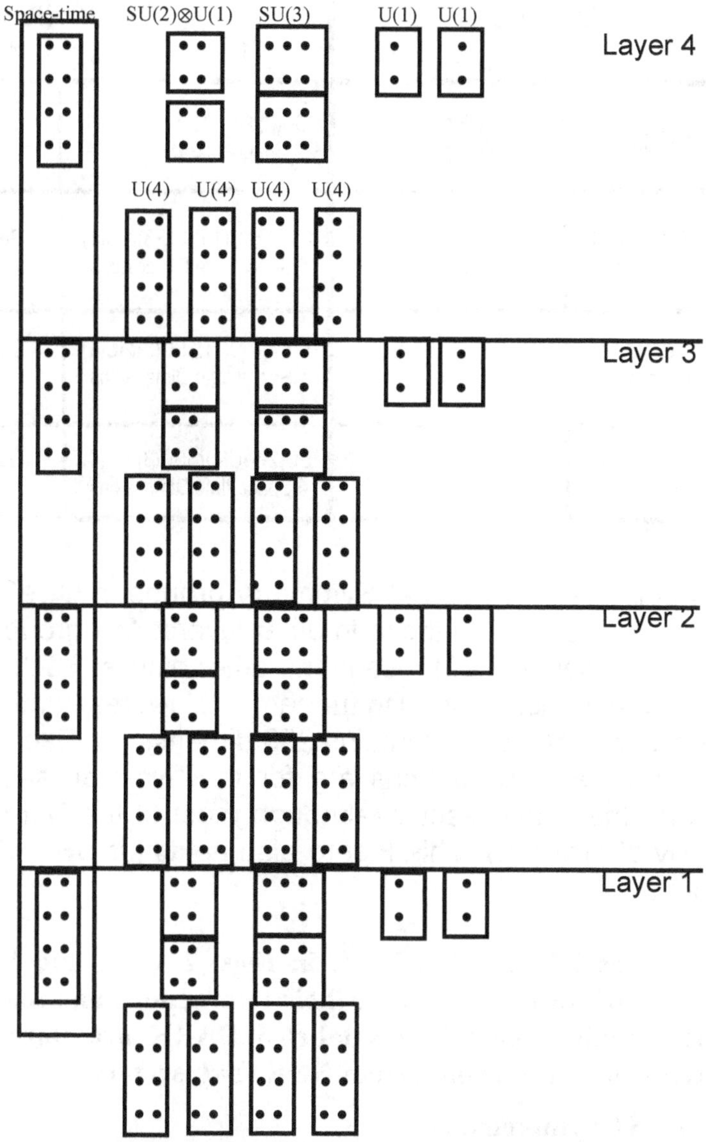

Figure 2.2. The four layers of QUeST internal symmetry groups (and space-time) for 32 dimension complex quaternion space. This is a changed Fig. 1-A.3 (U(2) → *U(1)⊗U(1))*. It is done to support the symmetry breakdown pattern above. Note: each row has an 8 • complex quaternion. Note the left column of blocks combine to specify a 4 dimension complex quaternion space-time. Note each layer requires 64 dimensions.

	NORMAL		DARK	
	4	4	4	4
4	SU(2)⊗U(1)⊗SU(3)⊗U(1) 4 Space-time Dimensions	Generation + Layer Groups	SU(2)⊗U(1)⊗SU(3)⊗U(1) 4 Space-time Dimensions	Generation + Layer Groups
4	SU(2)⊗U(1)⊗SU(3)⊗U(1) 4 Space-time Dimensions	Generation + Layer Groups	SU(2)⊗U(1)⊗SU(3)⊗U(1) 4 Space-time Dimensions	Generation + Layer Groups
4	SU(2)⊗U(1)⊗SU(3)⊗U(1) 4 Space-time Dimensions	Generation + Layer Groups	SU(2)⊗U(1)⊗SU(3)⊗U(1) 4 Space-time Dimensions	Generation + Layer Groups
4	SU(2)⊗U(1)⊗SU(3)⊗U(1) 4 Space-time Dimensions	Generation + Layer Groups	SU(2)⊗U(1)⊗SU(3)⊗U(1) 4 Space-time Dimensions	Generation + Layer Groups

Figure 2.3. Four layers of Internal Symmetry groups in QUeST. The groups in each layer are independent of those in other layers. The groups in each block of each layer are independent of those in the other blocks. Each block contains 16 dimensions. The dimensions furnish fundamental representations for the groups listed. The entire set of blocks contains 256 dimensions. Each layer contains 64 dimensions. The first two columns are for the "Normal" sector. The last two columns are for the "Dark" sector (although most of the Normal sector is Dark observationally at present.) This Figure also holds for UST with the addition of Fermion groups.

Note: The columns labeled NORMAL in Figs. 2.3, 2.4 and later include three layers that are Dark in the sense that their interactions and matter are as yet not found experimentally. The columns labeled DARK are Dark, now and in the future. DARK sectors are distinct from NORMAL sectors.

2.2 QUeST → UST Symmetries

The 16 tiles of Fig. 2.3 contain various symmetry groups and space-time. They are multiple copies of the same internal symmetries. However their coupling constants, particle masses, and symmetry breaking parameters will very likely differ.

We are familiar with part of the one known tile: three known generations of Normal fermions, the SU(2)⊗U(1)⊗SU(3) ElectroWeak and Strong interactions, and some masses, couplings and symmetry breaking parameters. The space-time is also known. The other tiles are completely unknown although our Quaternion-based theory suggests their likely form.

We now turn to mapping the above features of QUeST to UST. The QUeST symmetries listed in columns 1 and 3 of Fig. 2.3 do not appear in UST. We augment

UST, which is derived from axioms, by including these U(1) symmetries. The space-time coordinates are separate in UST. UST has 3+1 dimensions separate from internal symmetry; QUeST unites all dimensions for internal symmetry and space-time—one of its motivating goals.

Otherwise the pattern of QUeST symmetries matches the pattern of UST. We used this correspondence to determine QUeST symmetries of columns 1 and 3 in Fig. 2.3 while adding a U(1) Fermion group (discussed below) to each of their tiles. Further we use the UST symmetries of columns 2 and 4 of Fig. 2.4 to determine the corresponding columns of QUeST. These choices are appropriate since the QUeST dimension array does not inherently determine the physical nature of QUeST groups. Instead we see a *Correspondence Principle* where UST determines the physics of the QUeST array. Consistency of the QUeST array groups with the UST groups is required. This section demonstrates consistency if Fermion U(1) groups are added to UST. Satisfyingly, the U(1) groups have a natural physical interpretation in UST.

	NORMAL		DARK	
4	SU(2)⊗U(1)⊗SU(3)	Generation + Layer Groups	SU(2)⊗U(1)⊗SU(3)	Generation + Layer Groups
4	SU(2)⊗U(1)⊗SU(3)	Generation + Layer Groups	SU(2)⊗U(1)⊗SU(3)	Generation + Layer Groups
4	SU(2)⊗U(1)⊗SU(3)	Generation + Layer Groups	SU(2)⊗U(1)⊗SU(3)	Generation + Layer Groups
4	SU(2)⊗U(1)⊗SU(3)	Generation + Layer Groups	SU(2)⊗U(1)⊗SU(3)	Generation + Layer Groups

Figure 2.4. Four layers of Internal Symmetry groups in UST from Blaha (2020c) and earlier books such as Blaha (2018e). The groups in each layer are independent of those in other layers. The groups in each block of each layer are independent of those in the other blocks. The first two columns are for the "Normal" sector. The last two columns are for the "Dark" sector (although most of the Normal sector is Dark observationally at present.) Note columns 1 and 3 of UST omit a U(1) group relative to Fig. 2.3. Also space-time is separate and not included with of the set of symmetries in UST. The UST tiles in columns 1 and 3 have 10 dimensions; the tiles in columns 2 and 4 have 16 dimensions.

We now consider the Generation, Layer and Fermion groups. These groups occur in UST. We map them back to QUeST using the previously mentioned Correspondence Principle..

2.3 U(4) Generation, U(4) Layer and U(1) Fermion Groups

The Generation, and Layer groups were introduced in UST. (See Blaha (2018e), 2020c) and earlier books.) Their basis is in number conservation laws such as Baryon conservation. The Fermion group, now introduced, is also based on conservation laws:

We now describe the Generation and Layer groups briefly. Blaha (2020c) provides a more detailed description as does earlier books. The Fermion group is described later.

2.4 The Generation Group

We define two particle number operators for normal up-quark particles and down-quark particles, B_{uq} and B_{dq}. Similarly we define two particle number operators for normal species "e" (electron) particles and species "ν" particles, B_e and B_ν. Similarly we define Dark matter equivalents:[32] B_{De}, $B_{D\nu}$, B_{Duq}, and B_{Ddq}.

In the absence of interactions these fermion particle number operators are conserved. Each set are "diagonal" operators within a U(4) group. Thus we have a normal U(4) Generation Group and a Dark U(4) Generation group *for each layer.*.

On this basis we find there are four generations of each species in the normal and in the Dark matter sectors since U(4) groups have 4-dimension fundamental representations. One generation of normal fermions with large masses has not as yet been found in the lowest layer; three generations of normal fermions have been found. .

The broken symmetry gauge vector bosons of the Generation Group also have large masses. If the conservation of the fermion particle numbers is broken then we view it as a consequence of Generation Group symmetry breaking.

The Generation group of each layer generates interactions between the fermions of that layer. See Fig. 1-A.5 for the interaction between fermions generated by the Generation group of layer 4.

2.5 The Layer Group

The set of particle number operators can be extended if we take account of the fourfold fermion generations. To further refine the set of particle number operators we temporarily neglect all interactions that would violate conservation laws for the set.

We therefore subdivide the above particle number set into four particle numbers per generation. For the ith generation we define

L_{ie} – The "e" species particle number for the ith generation
$L_{i\nu}$ – The ν species particle number for the ith generation
L_{iuq} – The up-quark species particle number for the ith generation
L_{idq} – The down-quark species particle number for the ith generation

---

[32] By analogy, we assume that there are four species of Dark matter: charged Dark leptons, neutral Dark leptons, Dark up-type quarks, and Dark down-type quarks. Thus we are led to the Dark particle numbers: Dark Baryon Numbers, and Dark Lepton Numbers shown above.

L_{iDe} – The Dark "e" species particle number for the ith generation
$L_{iD\nu}$ – The Dark ν species particle number for the ith generation
L_{iDuq} – The Dark up-quark species particle number for the ith generation
L_{iDdq} – Dark down-quark species particle number for the ith generation

for each generation i = 1, 2, 3, 4. Individual fermions have positive $L_{ia} = +1$ values and anti-fermions have negative $L_{ia} = -1$ values for species a = 1, 2, 3, 4 (with the three color subspecies of quarks treated as part of one species.)

At this point we have four particle number operators for each generation. We define a group framework for each set of particle numbers. The simplest way is to assume that each generation consists of the four layers with the particles in each generation in a U(4) fundamental representation.[33] Then each generation has a U(4) Layer group with the generation's four number operators (above) as its diagonal operators. We call this group the Layer Group of the ith generation L_{ia}. With four generations we obtain four U(4) Layer groups for normal matter. In addition there are four U(4) Dark Layer groups. See Fig. 2.4.

The consequence of this expansion of particle numbers and groups is that the set of fermions increases fourfold. We now have four layers, with each having four generations. Experimentally, we know of three generations of fermions—the lowest generations of the lowest level. The remaining generation and the three additional levels of fermions are of much higher mass and yet to be found.

See Blaha (2020c) and (2018e) for a detailed discussion of the Layer Groups. We note in passing that the symmetries of these number operators are badly broken. Yet the underlying group structure remains.

Note each Layer group provides an interaction among layers: the ith Layer group provides an interaction between the four i^{th} generations. See Fig. 1-A.5 for the four Layer groups' interactions between fermions for the four generations. Layer groups provide the only interactions between layers.

2.6 The Fermion Groups

For each layer of the NORMAL sector there is a U(1) Fermion group.[34] There is a similar Fermion group for each layer of the DARK sector. See Fig. 2.3. These groups were introduced in QUeST. We have added them to UST since they have natural roles.

We will attribute them to fermion numbers conservation[35] for each layer and for NORMAL and DARK separately—one conservation law for total fermion number in each tile of columns 1 and 3 of Fig. 2.3. *In each of the columns 1 and 3 tiles fermion number equals the sum of Baryon number and Lepton number. It appears to be conserved in particle interactions. It can be conserved even if Baryon number*

---

[33] See Fig. 2.3 for a depiction of the "splitting" of fermions: first into generations, then into layers.

[34] Earlier versions of QUeST this year had a broken U(2) Dark symmetry group that gave interactions between Normal and Dark particles. Here we replace this group with U(1)⊗U(1) Fermion groups due to a tile by tile conserved fermion number. Appendix 1-A uses the Dark U(2) group. The conent of that appendix can be modified to Fermion groups by replacing each Dark U(2) with U(1)⊗U(1).

[35] Fermion number conservation differs from charge conservation since it applies to both charged and uncharged fermions.

conservation and Lepton number conservation were violated. The sum of all NORMAL Fermion numbers F_{tot} is also conserved. The sum of all DARK Fermion numbers F_{Dtot} is conserved as well.

For the NORMAL and DARK sectors of each layer of fermions we define an additive Fermion number N_f that equals the number of fermions of that layer minus the number of antifermions of that layer. In an interaction where fermion number is conserved, the input N_f equals the output N_f. This conservation law, which resembles the baryon number conservation law, is broken by Layer group interactions. Thus the broken Fermion groups, like the Generation and Layer groups, are based on a conserved number. The gauge fields of the Fermion groups will be denoted Y^μ and $Y_D{}^\mu$ for the NORMAL and DARK sectors of each layer respectively.

UST with Fermion groups added follows from QUeST.

Exercise: Replace the Dark U(2) group in Appendix 1-A with U(1)⊗U(1) Fermion groups.

2.7 Particle Spectrums of QUeST: Fermions, and Vector Bosons

In this section we show that QUeST and UST have the same vector boson and fermion particle spectrums.

2.7.1 Vector Bosons

An examination of Figs. 2.2 and 2.3, which now apply to both QUeST and UST, directly yields the spectrum of gauge vector bosons. This spectrum includes the known vector bosons of the Standard Model. The list of vector boson groups and interactions *for one of the four layers* is:

<u>NORMAL Gauge Groups</u>
SU(2)⊗U(1)⊗SU(3)⊗U(1)
Generation Group U(4)
Layer Group U(4)
<u>DARK Gauge Groups</u>
SU(2)⊗U(1)⊗SU(3)⊗U(1)
Generation Group U(4)
Layer Group U(4)

2.7.2 Fermions

Having the spectrum of gauge vector bosons the fundamental fermion spectrum also follows directly from the dimensions of the fundamental group representations. See Fig. 1-A.5 for the QUeST and UST fermion spectrums.

The basis for the fermion spectrum is twofold:

1. There are four species of fermions due to the boosts of the Complex Lorentz group as shown in Blaha (2020c) and earlier books. These species are charged lepton, neutral lepton, up-type quark, and down-type quark.[36]

[36] While this was shown as early as 2007 by the author, it became very evident as a general form when it was shown to hold in the UTMOST megaverse earlier in this year.

2. The leptons were seen to be singlets. The quarks were seen to be Color triplets.

3. With these points in mind there are 8 fermions for each of the eight appearances of SU(2)⊗U(1)⊗SU(3)⊗U(1) in Fig. 2.3. See Fig. 1-A.5.

3. Fundamental QUeST-UST Axioms

For a number of years the author has been arguing for an axiomatic derivation of the theory of elementary particles and gravitation. The axioms for UST generated that theory.[37] The axioms are listed in section 3.2. A preliminary basis for the axioms appears in section 3.1. An extended set of axioms for the combined QUeST-UST theory appears in Fig. 3.3.

3.1 Fundamental Prerequisites for a Fundamental Theory of Physics

We can list fundamental prerequisites based on a general knowledge of the necessary nature of a fundamental theory of Physics. This approach presumes a general knowledge of the theory that we wish to construct illustrating the maxim, "Our ends determine our beginnings."

A. A time variable must exist that may have various forms.

B. We wish to have a dynamical fundamental theory that evolves in time. Thus there must be a mechanism(s) that allow dynamical processes to exist that may, or may not, run in parallel.

C. Multiple parallel processes can execute. This requires a space with at least 4 dimensions.

D. There must be a space with a coordinate system(s), and distance measure, within which processes can execute.

E. There must be particles upon which dynamical processes execute.

F. There must be a space of functionals that support the creation of particle states and help determine their properties. The particle functional space frees particles from a complete dependence on coordinate space and supports instantaneous quantum entangled processes.

G. There must be a space of 'waves' of free field fourier expansions for all the fundamental particles absent interactions..

[37] See Blaha 2018e).

H. There must be an order in the 'created' dynamical theory that is embodied in a form of a computational language[38] with a Chomsky-like *Grammar* using a finite set of terminal and non-terminal *symbols* that constitutes an alphabet (vocabulary).[39] The ordering in the form of a language with grammar *Production Rules* ensures the consistency of the generated theory.

I. Creation should opt for Vitamorphic[40] universes that support life in some form. Recent studies have shown that evolution favors the development of increasingly intelligent life. Thus the ultimate appearance of intelligent life at places within universes appears to be natural—making the *Anthropic Principle* an evolutionary consequence[41] of the *Vitamorphic Principle.*

These prerequisites would seem to be necessary and sufficient for the specification of primitive terms and axioms for a fundamental theory.

3.1.1 Particles, Quantum Field Theory, Quaternions, Higgs Particles

Simple questions arise in deriving UST from the above axioms:

1. Why particles and not a continuum?
   Particles are the simplest way of creating an interacting theory. The alternative of a continuous range of matter with masses is far more complex.

2. Why Quantum Field Theory?
   Quantum Field Theory is well adapted to describe particle dynamics. It is based on discrete sets of particles. Quantum Field Theory provides a framework for particle interactions.

---

[38] The possibility that the universe is one enormous Word was explored in *Cosmos and Consciousness* (Blaha (2003)) in physical, philosophical, and religious contexts. A few years ago around 2012 the author found a book with a similar title by R. M. Bucke published in 1901 entitled *Cosmic Consciousness* on the evolution of Man to a new level of consciousness. The content of this book is inrelated to Blaha (1998) – first edition - and (2003) as well as Blaha's other books.

[39] Particle Computer Languages are described in Blaha (2005b) and (2005c) as well later in this volume and in other books by the author.

[40] The *Vitamorphic Principle* states that universes should support some form of life realizing that there are many varieties of life and borderline forms of life. A 'tight' definition of life has not been satisfactorily constructed. There are many borderline entities that may or may not be called life. We take 'Vitamorphic' to mean 'life enabling' in English. Vitamorphism is not a concept without meaning—a universe (Megaverse) consisting of only inert matter without energy present would be non-Vitamorphic. The Anthropic Principle, briefly put, states that intelligent human-like life should exist.

[41] One can well wonder whether the emergence and dominance of Mankind has eliminated the possibility of the emergence of other intelligent species on earth from the many semi-intelligent species that exist now and in the past.

3.2 UST Axioms

The set of axioms is

PARTICLE AXIOMS

1. All matter and energy is composed of particles.
2. Each fundamental particle has a physico-logic structure within it that we designate its core.
3. Particles form an alphabet with a finite number of characters and combine in ways specified by the quantum probabilistic production rules of a quantum computational grammar.[42]
4. A core is a particle functional that combines with a free field fourier coordinate expansion in an inner product to produce a free second quantized particle field.
5. There is a 4-dimensional space of particle functionals, called *particle functional space*, with the distance measure, eq. 1-A.1 below, specifying the transformation group of particle functionals.
6. Particle functional space consists of a single point.
7. The core of a fermion functional is called a *qube*. Fundamental bosons have a core consisting of a boson functional called a *quba*.
8. Qubes have a bare mass. Qubas have zero mass.

SPACE AXIOMS

9. The dimensions of a coordinate space-time are determined by the number of fundamental[43] interactions, and the requirement that all parallel processes, with parts perhaps separated by distances, can occur synchronously.
10. Spatial coordinates are inherently complex-valued.
11. Space has one complex-valued component that plays the role of time. Physical phenomena dynamically evolve based on the time variable.
12. The infinitesimal distance ds between two space-time points is given by

$$ds^2 = dt^2 - d\mathbf{x}^2$$

    where $d\mathbf{x}$ is a vector of the spatial coordinates. Transformations between coordinate systems preserve the value of ds and define a transformation group. (The Complex Lorentz Group)
13. Physically acceptable reference frames have real-valued coordinates. These coordinates can be obtained by group transformations from complex-valued coordinate systems. Physical space-time measurements are made in a real-valued coordinate system.
14. The speed of light is the same in all reference frames.
15. Free fundamental leptons must have a real-valued energy.

---

[42] See Blaha (2005b).

[43] Interactions that would exist in the absence of fermion particles.

16. Gravity may cause space-time to be curved. (Complex General Coordinate transformations[44])

DYNAMICS AXIOMS

17. The complete theory has a lagrangian formulation. If the lagrangian is truncated to quadratic form (with interactions set to zero) then symmetries appear that are the source of particle symmetry groups that persist with broken symmetry after interactions are reintroduced. The lagrangian specifies a set of production rules of a type 0 Chomsky language generalized to include production rules for the generation of all strings of symbols (particles) from any strings of symbols (including the *head symbol*.)[45]
18. The lagrangian of the theory must be invariant under coordinate system transformations.
19. Dynamical particle equations must be covariant under group transformations.
20. All interactions have a local Yang-Mills gauge theory formulation.
21. The vector bosons, and the interactions among them, are determined by terms in complete lagrangian, some of whose parts are obtained from the Riemann-Christoffel Curvature Tensor.

QUANTIZATION AXIOMS

22. All fields must be canonically quantized.
23. Fermion and Boson vacuums can be defined that are valid in all coordinate systems.
24. The number of particles in an asymptotic state of any given type is invariant in all reference frames.
25. Quantum processes starting in an initial quantum state, with parts separated by a distance after a time, can have the parts synchronously change each other instantaneously. (Quantum Entanglement)

3.2.1 The Derivation of the Unified SuperStandard Theory

The derivation of the Unified SuperStandard Theory has been a multi-year process undertaken by the author. Much of the derivation appears in Blaha (2015a), (2016f), (2017b), (2017c), (2017d), (2018a), (2018b), and (2018c). Earlier work, upon which these books are based, is referenced in these books.

---

[44] If the metric tensor of space-time is analogous to one of the metric tensors of the superfluid phases of ^{3}He, then space-time might have several metric tensors in 'various regions.' If the space-time metric tensor is analogous to the ^{3}He-B superfluid phase metric tensor, which has an effective gravity with a complex metric tensor, the space-time metric tensor would be the familiar one of General Relativity. However if the space-time metric tensor is analogous to the metric tensor of superfluid ^{3}He-A, which exists at higher pressure and temperature, then the space-time metric tensor might be similar to the Penrose twistor theory metric tensor. In this case the corresponding General Relativity may have a twistor-like metric tensor: perhaps in the early universe, and/or inside black holes, and/or in small universes with higher pressure and temperature than our universe. We will assume the conventional metric for Complex Special and General Relativity.

[45] Chapter 8 of Blaha (2018b) discusses computational languages for particles in detail.

3.3 New Deeper Axioms for QUeST-UST

The basis for our QUeST-UST approach is the trend of types of coordinates:

Real → complex → quaternion → Complex quaternion → octonion → Complex octonion

The Standard Model is based on real-valued coordinates. The Unified SuperStandard Theory, and its precursors by the author, was based on complex coordinates (specifically Complex Lorentz group transformations). The author has developed biquaternion (complex quaternion), and bioctonion (complex octonion), SuperStandard Models in Blaha (2020a) and Blaha (2020b) with features that *directly lead to the Unified SuperStandard Theory upon restriction to real-valued coordinates.*

We create a Unified SuperStandard Theory in 32 complex quaternion coordinates (QUeST) for our universe, and extracted the Unified SuperStandard Theory (UST)..These theories were outlined in Blaha (2020a) and (2020b).

These new theories mesh well with the universe. Upon restriction to 3 + 1 real-valued coordinates QUeST becomes the Unified SuperStandard Theory of our universe described in earlier books and later in this book. Upon restriction to 7 + 1 real-valued coordinates UTMOST becomes the theory of the Megaverse described in earlier books and later in this book

AXIOMS

1. A biquaternion space is the basic space of our universe. A bioctonion space is the basic space of the Megaverse. These spaces factorize into a coordinate space-time and an internal symmetry space-time.

2. Physical processes can execute in parallel.

3. Matter and energy are particulate.

4. Space--times are locally Lorentzian.

5. All calculations are finite.

6. Particle theory can be defined in any curved space-time.

7. Each particle has a wave function determined by a functional inner product defining the particle state. The functionals form a set without a distance measure.

3.4 General Implications of the QUeST-UST Axioms

In this section we describe some of the implications of each of the axioms.

1. Biquaternion space is the basic space of our universe. Bioctonion space is the basic space of the Megaverse. These spaces each factorize into a coordinate space-time and an internal symmetry space.

The factorization into a space-time and an internal symmetry space must be a form of spontaneous symmetry breaking of yet unknown origin. It appears to be related to a breakdown of the vacuum.

2. Physical processes can execute in parallel.

Physical processes are known to be able to execute in parallel at any distance of separation. As Fant has shown parallel execution requires a minimal number of dimensions: 4. Consequently the dimension of space-time must be 4 or greater. The biquaternion space-time of QUeST is 4-dimensional allowing parallel process execution.

The bioctonion space-time of MOST is 8-dimensional and also allows parallel process execution. The choice of eight dimensions is natural since it allows 4-dimensional universes within it. It also has a form that allows a clean formulation. Lastly, as will be seen later, it conforms to the pattern of interplay between Lorentz symmetry and internal symmetry found in the Unified SuperStandard Theory..

3. Matter and energy are particulate.

The most direct method of specifying a theory of matter and energy is through the Use of Quantum Field Theory. Thus Quantum Field Theory is implied.

4. Complex Space-times are locally Lorentzian.

A locally complex Lorentzian space-time leads to Complex General Relativity. In flat space-time Complex General Relativity becomes Complex Lorentz group. (In point of fact the Complex Poincarė group follows.

5. All calculations are finite.

Given the need for Quantum Field Theory it becomes necessary to find a formulation that yields finite values for calculations in perturbation theory. The only approach that eliminates high energy divergences, and yet preserves the results found in perturbation theory calculations that agree with (primarily QED) experiments, is Two-Tier Quantum Field Theory. This is discussed in detail in earlier books starting in 2002. Thus only our Two-Tier formalism satisfies this axiom.

6. Particle theory can be defined in any curved space-time.

In the 1970s we developed a formalism that allows the definition of particle states in any space-time in such a way that its physical content is preserved when transformed to any coordinate system.[46] This PseudoQuantum Quantum Field Theory satisfies this axiom.

7. Each particle has a wave function determined by a functional inner product defining the particle state. The functionals form a set without a distance measure.

This axiom is satisfied by our formulation of quantum functionals in Blaha (2019f) and earlier books. Our formulation eliminates the superficial violation of the Theory of Relativity by "spooky" quantum entangled processes with parts separated by a physically "large" distance.

The seven axioms imply the Unified SuperStandard Theories and its deeper biquaternion and bioctonion hypercomplex formulations.

The following chapters 4 through 9 describe aspects of QUeST-UST. Then the origin of UTMOST in BMOST is derived.

---

[46] S. Blaha, Il Nuovo Cimento **49A**, 35 (1979).

4. Enhanced Quantum Field Theory From Chapters 40 and 41 of Blaha (2020c)

This chapter contains chapters 40 and 41 of Blaha (2020c) describing Two-Tier coordinates and PseudoQuantum Field Theory. They resolve the problem of infinities in conventional quantum field theory. They also support quantization in any coordinate system as seen in Appendix 1-C as well as supporting the derivation of QUeST from BQUeST and UTMOST from BMOST.

40. Two-Tier Coordinates

Originally Two-Tier coordinates were developed by this author to remove infinities that appear in perturbation theory calculations. We showed that the quantum smeared coordinates of Two-Tier Quantum Field Theory succeeded in removing all ultra-violet infinities in perturbation theory including the fermion triangle infinities. Remarkably the high precision, low energy[47] predictions of QED remained true in Two-Tier QED and thus remained consistent with experiment to a hitherto unsurpassed level of accuracy. 'Low' energy predictions in other quantum field theories also remained unchanged. At high energies, Two-Tier perturbation theory results are finite and consequently all ultra-violet infinities, to any order in perturbation theory, in *any number of space-time dimensions* were eliminated.

In addition to removing perturbation theory infinities Two-Tier coordinates enable us to define finite theories of Quantum Gravity and 'non-renormalizable' quantum field theories based on polynomial lagrangians, to tame vacuum fluctuations, to eliminate infinities associated with the Big Bang, and possibly to generate the explosive growth of the universe in its role as Dark Energy.[48]

Two-Tier Quantum Field Theory is established on the most fundamental level.

40.1 Two-Tier Features in 4-Dimensional Space-Time

Two-Tier Quantum Field Theory,[49] which was based on a new method[50] in the Calculus of Variations, uses two sets of fields to introduce quantum coordinates. We shall consider this technique for the specific case of a massless vector field $V^i(y)$ analogous to the electromagnetic field.

In 4-dimensional space-time the massless vector field has the form $Y^\mu(y)$ where the index μ ranges from 0 through 3. The X^μ coordinate system, where it appears, has a c-number real part and a q-number imaginary part. Thus particle fields which are

---

[47] Relative to a mass scale that was perhaps of the order of the Planck mass.

[48] See Blaha (2017b) and earlier books for details. This section is basically a summary of some features.

[49] See Blaha (2005a), and Blaha (2002), for discussions of this new method to eliminate infinities in quantum field theory calculations.

[50] Blaha (2005a) describes our method for the composition of extrema in some detail.

normally defined on four-dimensional real space-time will now be defined on a complex four-dimensional space-time where four imaginary dimensions will appear as *Quantum Dimensions* embodied in a vector quantum field $Y^\mu(y)$:

$$X^\mu(y) = y^\mu + i\, Y^\mu(y)/M_c^2$$

where M_c is an extremely large mass of the order of the Planck mass or perhaps much larger.

The $Y^\mu(y)$ field is a function of the subspace y coordinates. The real part of the space-time dimensions will be taken to be the space of real-valued y coordinates.[51]

The imaginary part of space-time coordinates is the a massless $Y^\mu(y)$ vector quantum field that is suppressed further by a very large mass scale – perhaps of the order of the Planck mass – that reduces the imaginary Quantum Dimensions to the infinitesimal except at large momenta. The effects of Quantum Dimensions only become appreciable in quantum field theory at energies of the order of M_c. At these energies exponential Gaussian factors in each particle (and ghost) propagator are generated by the Quantum Dimensions and serve to make perturbation theory calculations ultra-violet finite – including calculations in Quantum Gravity.

The formalism introduces a new form of interaction that does not have the form of the simple polynomial interactions that have hitherto dominated quantum field theories. This form of interaction takes place via the composition of quantum fields and can be called a *Dimensional Interaction,* or an *Interdimensional Interaction,* since it affects particle behavior through Quantum Dimensions.

The basic ansatz of the Two-Tier formalism is to replace every appearance of a coordinate x in a quantum field with the variable

$$x^\mu \rightarrow X^\mu = (y^0, \mathbf{y} + \mathbf{Y}(y^0, \mathbf{y})/M_c^2)$$

where $\mathbf{Y}(y^0, \mathbf{y})$ is the spatial part of a free massless vector field with features that are identical to the free QED field in the Radiation gauge.

Then one finds that the momentum space free field Feynman propagators G(k) of all particles acquires a Gaussian factor exp(h(k)):

$$G(k) \rightarrow G(k)\, \exp(h(k))$$

so that all perturbation theory diagrams are finite. The result is finite perturbative results for all calculations to any order in perturbation theory. Blaha (2005a) shows that Two-Tier theories are finite, Poincare covariant, and unitary. (See Blaha (2005a), chapter 5, for a complete discussion.)

---

[51] In a deeper theory the real part might also be a quantum field that undergoes a condensation to generate c-number coordinates. We will not consider this possibility in this book.

40.2 Simple Two-Tier X^μ Formalism

In this subsection we will describe the basic Two-Tier formalism. Taking the lagrangian described in Blaha (2005a):[52]

$$\mathscr{L}(y) = \mathscr{L}_F(X^\mu(y))J + \mathscr{L}_C(X^\mu(y), \partial X^\mu(y)/\partial y^\nu, y) \tag{40.1}$$

where

$$X^\mu(y) = y^\mu + i\, Y^\mu(y)/M_c^2 \tag{40.2}$$

with M_C being a large mass scale, $Y_\mu(y)$ a vector quantum field, and where J is the absolute value of the Jacobian of the transformation from X to y coordinates:

$$J = |\partial(X)/\partial(y)|$$

The lagrangian term $\mathscr{L}_C$ is

$$\mathscr{L}_C = +\tfrac{1}{4}\, M_c^4 F^{\mu\nu}F_{\mu\nu}$$

with

$$\begin{aligned} F_{\mu\nu} &= \partial X_\mu/\partial y^\nu - \partial X_\nu/\partial y^\mu \\ &\equiv i\,(\partial Y_\mu/\partial y^\nu - \partial Y_\nu/\partial y^\mu)/M_c^2 \end{aligned} \tag{40.3}$$

The lagrangian term $\mathscr{L}_F(X^\mu(y))$ contains the terms for scalar, fermion and other gauge terms in general. The sign in $\mathscr{L}_C$ is not negative – contrary to the conventional electromagnetic Lagrangian. The reason for this difference is that the quantum field part of X^μ is imaginary. Thus $\mathscr{L}_C$ ends up having the correct sign after taking account of the factor of i in the field strength $F_{\mu\nu}$.

Defining

$$F_{Y\mu\nu} = (\partial Y_\mu/\partial y^\nu - \partial Y_\nu/\partial y^\mu)$$

we see the Lagrangian assumes the form of the conventional electromagnetic Lagrangian:

$$\mathscr{L}_C = -\tfrac{1}{4}\, F_Y^{\mu\nu}F_{Y\mu\nu}$$

The action of this theory has the form

$$I = \int d^4y\, \mathscr{L}(y)$$

40.3 Y^μ Gauge

The gauge invariance of the Lagrangian allows us to choose a convenient gauge. The gauge invariance of the full Lagrangian

$$\mathscr{L}_s = L_F(\phi(X), \partial\phi/\partial X^\mu)\, J + \mathscr{L}_C(X^\mu(y), \partial X^\mu(y)/\partial y^\nu)$$

[52] Eq. 7.1. See Appendix D for more detail.

is based on the standard gauge invariance of $\mathscr{L}_C$, and the gauge invariance of $J\mathscr{L}_F$ in the form of translational invariance

$$X^\mu(y) \to X^\mu(y) + \delta X^\mu(y)$$

for the special case of a translation of X with the form of a gauge transformation:

$$\delta X^\mu(y) = \partial\Lambda(y)/\partial y_\mu$$

In this case we find

$$\int d^4y\ \Lambda(y)\ \partial\ [\ J\,\partial/\partial X^\mu\ \mathscr{T}_{F\mu\nu}\ ]/\partial y_\nu\ = 0 \qquad (40.4)$$

after a partial integration and so we have the differential conservation law:

$$\partial\ [\ J\,\partial\mathscr{T}_{F\mu\nu}/\partial X^\mu]/\partial y_\nu\ = 0$$

since $\Lambda(y)$ is arbitrary. This conservation law is trivially obeyed:

$$\partial\mathscr{T}_{F\mu\nu}/\partial X^\mu = 0 \qquad (40.5)$$

Thus translational invariance in the $\mathscr{L}_F$ sector together with standard gauge invariance in the $\mathscr{L}_C$ sector automatically guarantees Y field gauge invariance of the total Lagrangian. We use the separate invariance of each term of

$$L = \int d^4y\ [\mathscr{L}_F\, J + \mathscr{L}_C\,] = \int d^4X\ \mathscr{L}_F\ + \int d^4y\ \mathscr{L}_C = L_F + L_C$$

under a constant translation $X^\mu \to X^\mu + \delta X^\mu$ where δX^μ is constant. Then we consider a position dependent translation/gauge transformation, which taken together with the above equation, establishes the invariance under the position dependent translation/gauge transformation.

An alternate approach that leads to the same result is to start with the particle part of the Lagrangian $\mathscr{L}_F$ rewritten to be invariant under general coordinate transformations, as it must, when we generalize to include General Relativity. Since position dependent translations are a form of general coordinate transformation the full theory must be invariant under position dependent translations due to invariance under general coordinate transformations.

Having established invariance under gauge transformations we now choose to use the most convenient gauge – the radiation gauge[53]:

$$\partial Y^i/\partial y^i = 0 \quad (40.6)$$

where i = 1, 2, 3, which, in the absence of external sources, allows us to set

$$Y^0 = 0$$

since Y^0 does not have a canonically conjugate momentum. A conventional treatment leads to the equal time commutation relations:

$$[Y^\mu(\mathbf{y}, y^0), Y^\nu(\mathbf{y}', y^0)] = [\pi^\mu(\mathbf{y}, y^0), \pi^\nu(\mathbf{y}', y^0)] = 0 \quad (40.7)$$

$$[\pi^j(\mathbf{y}, y^0), Y_k(\mathbf{y}', y^0)] = -i\,\delta^{tr}{}_{jk}(\mathbf{y} - \mathbf{y}')$$

(Note the locations of the j indexes above introduce a minus sign.) where

$$\pi^k = \partial\mathscr{L}_C/\partial Y_k'$$
$$\pi^0 = 0$$

$$\delta^{tr}{}_{jk}(\mathbf{y} - \mathbf{y}') = \int d^3k\, e^{i\,\mathbf{k}\cdot(\mathbf{y}-\mathbf{y}')}(\delta_{jk} - k_j k_k/\mathbf{k}^2)/(2\pi)^3$$

$$Y_k' = \partial Y_k/\partial y^0$$

The Radiation gauge reveals the two degrees of freedom that are present in the vector potential. The Fourier expansion of the vector potential is:

$$Y^i(y) = \int d^3k\, N_0(k) \sum_{\lambda=1}^{2} \varepsilon^i(k, \lambda)[a(k,\lambda)\, e^{-ik\cdot y} + a^\dagger(k,\lambda)\, e^{ik\cdot y}] \quad (40.8)$$

where

$$N_0(k) = [(2\pi)^3 2\omega_k]^{-1/2}$$

and (since m = 0)

$$\omega_k = (\mathbf{k}^2)^{1/2} = k^0$$

with $\vec{\epsilon}(k, \lambda)$ being the polarization unit vectors for $\lambda = 1,2$ and $k^\mu k_\mu = 0$.

The further development of this theory is described in Part 3 of Blaha (2005a).

[53] It is also possible to quantize using an indefinite metric that preserves manifest Lorentz covariance as was done by Gupta and Bleuler for the electromagnetic field. We will use the Gupta-Bleuler approach later to establish covariance under special relativity later. Now we opt for manifest positivity and use the radiation gauge.

40.4 Scalar Field Quantization Using X^μ

We will begin by considering the case of a scalar quantum field theory. We assume a real underlying y subspace. Since X^μ is a set of coordinates, we choose to define a scalar field ϕ as a function of X^μ, which, in turn, is a function of the y^ν coordinates. We will provisionally second quantize ϕ treating X^μ as c-number coordinates using a conventional approach.[54]

We assume a Lagrangian, with the momentum conjugate to ϕ:

$$\pi_\phi = \partial L_F / \partial\phi' \equiv \partial L_F / \partial(\partial\phi/\partial X^0) \tag{40.9}$$

Following the canonical quantization procedure, π and ϕ become hermitian operators with equal time ($X^0 = X^{0\prime}$) commutation rules:

$$\begin{aligned}[\phi(X), \phi(X')] &= [\pi_\phi(X), \pi_\phi(X')] = 0 \\ [\pi_\phi(X), \phi(X')] &= -i\,\delta^3(\mathbf{X} - \mathbf{X}')\end{aligned} \tag{40.10}$$

The standard Fourier expansion of the solution to the Klein-Gordon equation is:

$$\phi(X) = \int d^3p\, N_m(p)\, [a(p)\, e^{-ip\cdot X} + a^\dagger(p)\, e^{ip\cdot X}]$$

where

$$N_m(p) = [(2\pi)^3 2\omega_p]^{-1/2}$$

and

$$\omega_p = (\mathbf{p}^2 + m^2)^{1/2}$$

The commutation relations of the Fourier coefficient operators are:

$$\begin{aligned}[a(p), a^\dagger(p')] &= \delta^3(\mathbf{p} - \mathbf{p}') \\ [a^\dagger(p), a^\dagger(p')] &= [a(p), a(p')] = 0\end{aligned}$$

The reader will recognize the quantization procedure is formally identical to the standard canonical quantization procedure of a free scalar quantum field.

In the case of spin ½, spin 1 and spin 2 fields the standard quantization procedure *in terms of the X coordinate system* can also be followed in a way similar to the procedure in standard texts.

---

[54] Some texts are: Bogoliubov, N. N., Shirkov, D. V., *Introduction to the Theory of Quantized Fields* (Wiley-Interscience Publishers Inc., New York, 1959); Bjorken, J. D., Drell, S. D., *Relativistic Quantum Fields* (McGraw-Hill, New York, 1965); Huang, K., *Quarks, Leptons & Gauge Fields Second Edition* (World Scientific, River Edge, NJ, 1992); Kaku, M., *Quantum Field Theory* (Oxford University Press, New York, 1993); Weinberg, S., *The Quantum Theory of Fields* (Cambridge University Press, New York, 1995).

40.5 Scalar Feynman Propagators

The momentum space free field Feynman propagators G…(k) of all particles and ghosts in all Two-Tier Quantum Field Theories acquires a Gaussian factor exp(h(k)):

$$G\ldots(k) \rightarrow G\ldots(k)\exp(h(k))$$

so that all perturbation theory diagrams are finite. The result is a finite perturbative result in all calculations to any order in perturbation theory. Blaha (2005a) shows that Two-Tier theories are finite, Poincare covariant, and unitary.

An example of the Two-Tier effect on propagators is the case of the Two-Tier photon propagator. The Two-Tier photon propagator[55] is:

$$iD_F^{TT}(y_1 - y_2)_{\mu\nu} = -i\int \frac{d^4p\, e^{-ip\cdot z}\, g_{\mu\nu} R(\mathbf{p}, z)}{(2\pi)^4 (p^2 + i\varepsilon)} \qquad (40.11)$$

(since the imaginary parts can be taken to be zero: $y_{1i}^{\mu} - y_{2i}^{\mu} = 0$) where

$$z^{\mu} = y_{1r}^{\mu} - y_{2r}^{\mu}$$

$$R(\mathbf{p}, \mathbf{z}) = \exp[-p^ip^j\Delta_{Tij}(z)/M_c^4]$$

$$= \exp\{ -\mathbf{p}^2[A(v) + B(v)\cos^2\theta] / [4\pi^2M_c^4|\mathbf{z}|^2$$

with i, j = 1, 2, 3, and with $\Delta_{Tij}(z)$ being the commutator of the positive frequency part $Y^+_k(y)$ and the negative frequency part $Y^-_k(y)$ of $Y_k(y)$:

$$\Delta_{Tij}(z) = [Y^+_j(y_{1r}), Y^-_k(y_{2r})] = \int d^3k\, e^{ik\cdot(y_{1r} - y_{2r})}\, (\delta_{jk} - k_jk_k/\mathbf{k}^2)/[(2\pi)^3 2\omega_k] \qquad (40.12)$$

and

$$v = |z^0|/|\mathbf{z}|$$
$$A(v) = (1 - v^2)^{-1} + .5v\ln[(v-1)/(v+1)]$$
$$B(v) = v^2(1 - v^2)^{-1} - 1.5v\ln[(v-1)/(v+1)]$$
$$\mathbf{p}\cdot\mathbf{z} = |\mathbf{p}|\,|\mathbf{z}|\cos\theta$$

with $|\mathbf{p}|$ denoting the length of a spatial vector **p**, $|\mathbf{z}|$ denoting the length of a spatial vector **z**, and with $|z^0|$ being the absolute value of z^0.

The gaussian factors R(**p**, z) which appear in all Two-Tier propagators damp the large momentum behavior of all perturbation theory integrals producing a completely

[55] Blaha (2005a).

finite perturbation theory and yet give the usual results of perturbation theory at energies that are small compared to the mass scale M_c.

40.6 String-like Substructure of the Theory

Two-tier Quantum field Theory endows each particle with an extended structure that resembles the extended structure seen in boson string and Superstring theories. For example, Bailin (1994) use the operator[56]

$$V_\Lambda(k) = \int d^2\sigma \sqrt{-h}\, W_\Lambda(\tau, \sigma)\, e^{-ik\cdot X}$$

where X^μ is a quantized fourier expansion of the string fields (see eq. 7.22 of Bailin (1994)).

We note our X^μ coordinate-field has two transverse degrees of freedom due to gauge invariance, which also invites comparison to the boson string. A point of difference is that we have a well-defined quantum field theoretic formulation in conventional space-time that has the Standard Model as its "large distance" behavior thus introducing a note of reality that is not apparent in Superstring theories. We see that the interacting quantum field theories based on this approach also have good, finite, short distance behavior just as string theories.

The scalar, and other particles', Feynman propagators can be viewed as describing the propagation of a particle cloaked (accompanied) by a cloud of Y particles (which generates the $R(\mathbf{p}, y_1 - y_2)$ factor in the above propagator). If we examine the fourier transform of R(p, z) we see:

$$(2\pi)^4 R(\mathbf{p}, q) = \int d^4z\, e^{iq\cdot z} R(\mathbf{p}, z) = \int d^4z\, e^{iq\cdot z} \exp[-p^i p^j \Delta_{Tij}(z)/M_c^4] \qquad (40.13)$$

and we find

$$R(\mathbf{p},q) = \sum_{n=0}^{\infty} [i(2\pi M_c)^4]^{-n} (n!)^{-1} \prod_{j=1}^{n} [\int d^4k_j\, \theta(k_j^0)(\mathbf{p}^2-(\mathbf{p}\cdot\mathbf{k}_j)^2/\mathbf{k}_j^2)/(k_j^2 + i\varepsilon)]\, \delta^4(q - \Sigma\, k_r)$$

which can be interpreted as a "cloud" of Y particles dressing the "bare" particle propagator. (The apparent divergences for R(p, q) are an artifact of the expansion and the subsequent fourier transformation. They are not present in the $R(\mathbf{p}, y_1 - y_2)$ factor in the propagator. See Fig. 40.1 for the Feynman diagram of the Two-Tier 'cloaked' propagator as compared to the normal scalar particle Feynman propagator. The Two-Tier Feynman propagator is basically a conventional scalar propagator that is modified by coherent Y particle emission.[57]

---

[56] D. Bailin and A. Love, *Supersymmetric Gauge Field Theory and String Theory* (Institute of Physics Publishing, Philadelphia, PA, 1994) page 272.

[57] T. W. B. Kibble, Phys. Rev. **173**, 1527 (1968) and references therein. In particular see p. 1532 of Kibble's paper.

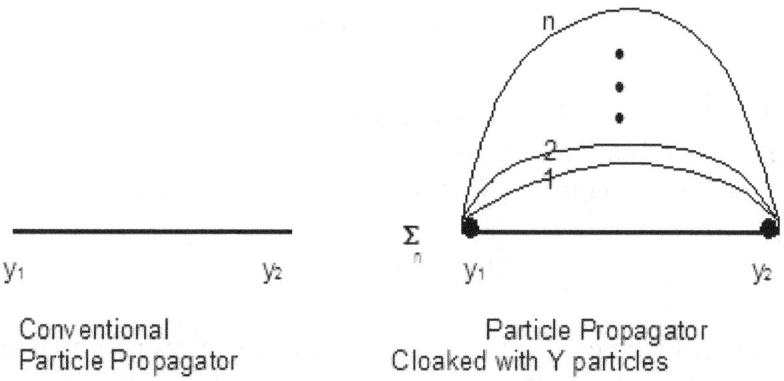

Figure 40.1. Feynman diagram for conventional and the nth diagram of a cloaked Two-Tier propagator.

We note that R(p, q) satisfies the convolution theorem:

$$\int d^4k \, R(\mathbf{p}, k)\, R(\mathbf{p}, q - k) = [R(\mathbf{p}, q)]^2$$

or

$$(2\pi)^4 \int d^4z \, e^{iq\cdot z} R(\mathbf{p}, z)\, R(\mathbf{p}, z) = [\int d^4z \, e^{iq\cdot z} R(\mathbf{p}, z)]^2 \tag{40.14}$$

The proof follows from the Binomial theorem.

40.7 Two-Tier Complexon Quantum Fields

In the case of the Complexon Standard Model we will need two variables X_r^μ and X_i^μ since we have complex spatial 3-coordinates. We define them similarly to the previous case:

$$X_r^\mu(y_r) = y_r^\mu + i\, Y_r^\mu(y_r)/M_c^2$$
$$X_i^\mu(y_i) = y_i^\mu + i\, Y_i^\mu(y_i)/M_c^2$$

where we choose the same mass scale for both the "real" and "imaginary" variables. The Two-Tier, single generation, version of the Complexon Standard Model then has an action of the form

$$I_{CSMtt} = \int dy^0 d^3y_r d^3y_i \, (\mathscr{L}_{CSM}(X_r^\mu(y_r), \mathbf{X}_i^k(y_i))J_2)|_{yi0 = 0,\, Yr0 = Yi0 = 0} + \tag{40.15}$$

$$+ \int dy_r^0 d^3y_r \, \mathscr{L}_C(X_r^\mu(y_r), \partial X_r^\mu(y_r)/\partial y_r^\nu, y_r) +$$

$$+ \int dy_i^0 d^3y_i \mathscr{L}_C(X_i^\mu(y_i), \partial X_i^\mu(y_i)/\partial y_i^\nu, y_i)$$

where the replacements

$$x^\mu \equiv x_r^\mu \rightarrow X_r^\mu(y_r)$$

$$x_i^k \rightarrow X_i^k(y_i)$$

for $\mu = 0, 1, 2, 3$ and $k = 1, 2, 3$ are made, followed by defining $y_r^0 = y^0$ and making a Complex Lorentz transformation to a frame where $y_i^0 = 0$. J_2 is the absolute value of the Jacobian of the transformation from (X_r, X_i) to (y_r, y_i) coordinates:

$$J_2 = |\partial(X_r, X_i)/\partial(y_r, y_i)|$$

We also choose gauges where $Y_r^0 = Y_i^0 = 0$. These types of transformations and gauge choices are discussed in detail in Blaha (2005a). The lagrangian terms $\mathscr{L}_C(X_r^\mu(y_r), \partial X_r^\mu(y_r)/\partial y_r^\nu, y_r)$ and $\mathscr{L}_C(X_i^\mu(y_i), \partial X_i^\mu(y_i)/\partial y_i^\nu, y_i)$ have the same form:

$$\mathscr{L}_C = +¼\, M_c^4 F^{\mu\nu} F_{\mu\nu} \tag{40.16}$$

with

$$F_{\mu\nu} = \partial X_\mu/\partial y^\nu - \partial X_\nu/\partial y^\mu$$
$$\equiv i\,(\partial Y_\mu/\partial y^\nu - \partial Y_\nu/\partial y^\mu)/M_c^2$$

or defining

$$F_{Y\mu\nu} = (\partial Y_\mu/\partial y^\nu - \partial Y_\nu/\partial y^\mu)$$

we see each lagrangian assumes the form of the conventional electromagnetic Lagrangian:

$$\mathscr{L}_C = -¼\, F_Y^{\mu\nu} F_{Y\mu\nu}$$

The lagrangian is supplemented with the following condition on all complexon fields $\Phi_{\ldots}$:

$$(\partial/\partial X_r^k(y_r))\,(\partial/\partial X_i^k(y_i))\Phi\ldots = 0 \tag{40.17}$$

summed over $k = 1, 2, 3$. Non-complexon fields $\Omega\ldots$ in our left-handed formulation satisfy the subsidiary condition:

$$\{(\partial/\partial X_r^k(y_r))(\partial/\partial X_i^k(y_i)) - [(\partial/\partial X_r^k(y_r))^2(\partial/\partial X_i^m(y_i))^2]^{½}\}\Omega\ldots = 0 \tag{40.18}$$

summed over $k = 1, 2, 3$ and over $m = 1, 2, 3$ separately in each of the two terms.

40.8 Complexon Feynman Propagator

In the case of complexons, the Two-Tier Feynman propagator differs from the non-complexon case by having an integration over imaginary spatial 3-momenta, a derivative of a delta function embodying the orthogonality of the real and imaginary 3-momenta, and two factors of R(**p**, z): one factor being R($\mathbf{p}_r$, z_r) and the other factor being R($\mathbf{p}_i$, z_i) (where the time components $z_r^0 = z^0$ and $z_i^0 = 0$ since there is only one real time coordinate[58]) thus providing large momentum convergence for both real and imaginary 3-momentum integrations.

For a normal scalar particle the Feynman propagator is:

$$i\Delta_{CTF}(x-y) = \theta(x^+ - y^+)\langle 0|\phi_{CT}(x)\,\phi_{CT}(y)|0\rangle + \theta(y^+ - x^+)\langle 0|\phi_{CT}(y)\phi_{CT}(x)|0\rangle$$
$$= i\int d^4p_r d^3p_i (2\pi)^{-7}\delta'(\mathbf{p}_r\cdot\mathbf{p}_i/m^2)e^{-ip^+(x^- - y^-) - ip^-(x^+ - y^+) + i\mathbf{p}_\perp\cdot(\mathbf{x}_\perp - \mathbf{y}_\perp) - i\mathbf{p}_i\cdot(\mathbf{x}_i - \mathbf{y}_i)}/(p^2 + m^2 + i\varepsilon) \quad (40.19)$$

in conventional quantum field theory.

In the case of Two-Tier quantum field a scalar *complexon* particle has the Feynman propagator

$$i\Delta_{CTFtt}(x-y) = i\int d^4p_r d^3p_i (2\pi)^{-7}\delta'(\mathbf{p}_r\cdot\mathbf{p}_i/m^2)\, R(\mathbf{p}_r, z_r)R(\mathbf{p}_i, z_i)\cdot \quad (40.20)$$
$$\cdot e^{-ip^+(x^- - y^-) - ip^-(x^+ - y^+) + i\mathbf{p}_\perp\cdot(\mathbf{x}_\perp - \mathbf{y}_\perp) - i\mathbf{p}_i\cdot(\mathbf{x}_i - \mathbf{y}_i)}/(p^2 - m^2 + i\varepsilon)$$

where the time components $z_r^0 = z^0$ and $z_i^0 = 0$ since there is only one time coordinate, where R(**p**, z) is given in the previous subsection, and where $p^2. = p^{0\,2} - p_r^{\,2} + p_i^{\,2}$.

Propagators for other types of particles are similarly modified in the Two-Tier formalism (See Blaha 2005a).

40.9 Vacuum Fluctuations

While the expectation value of a *conventional* free scalar field $\phi_{conv}(x)$ is zero in a conventional quantum field theory:

$$\langle 0|\phi_{conv}(x)|0\rangle = 0 \quad (40.21)$$

the vacuum fluctuations of *conventional* scalar quantum field theory are quadratically divergent:

$$\langle 0|\phi_{conv}(x)\phi_{conv}(x)|0\rangle = \int d^3p/[(2\pi)^3 2\omega_p] \quad (40.22)$$

In "Two-Tier" quantum field theory we find the vacuum expectation value of a free field is zero *and the expectation value of the square of the field is also zero:*

[58] We can arrange for $z_i^0 = 0$ by making a Complex Lorentz transformation to an inertial frame where z is real.

$$<0|\phi(X)\phi(X)|0> = \int d^3p\, e^{-p^ip^j\Delta_{Tij}(0)/Mc^4}/[(2\pi)^3 2\omega_p] = 0$$

since the exponential factor in the integral is $-\infty$. The exponent contains

$$\Delta_{Tij}(z) = \int d^3k\, e^{-ik\cdot z}\, (\delta_{ij} - k_ik_j/\mathbf{k}^2)/[(2\pi)^3 2\omega_k] \qquad (40.23)$$

where "T" is for "Two-Tier". Thus *vacuum fluctuations are zero in Two-Tier quantum field theory*. Correspondingly, we will see that renormalization constants are finite in the Two-Tier versions of QED, Electroweak Theory, the Standard Model and Quantum Gravity. See Blaha (2017b) and references therein for more details.

40.10 Time Intervals in General Relativity

Wigner[59] has studied the measurement of time intervals in General Relativity and sees a problem in the measurement of extremely short intervals. According to Wigner, the measurement of a time interval in a region of space requires the measurement of the length of time required for an event to happen. The measurement requires an accurate clock. But the accuracy of the clock is limited by the energy-time uncertainty relation:

$$\Delta E \Delta t \geq \hbar \qquad (40.24)$$

Thus the uncertainty in the clock's time measurement is related to the uncertainty in the clock's energy which is, in turn, related to the uncertainty in the clock's mass:

$$\Delta E = (\Delta m)c^2$$

To obtain "infinite" accuracy the uncertainty (fluctuations) in the clock's mass must be infinite and thus the clock's mass must be infinite. Infinite fluctuations in the clock's mass will produce corresponding infinite fluctuations in the gravitational field.

$$\Delta h \propto \Delta E \qquad \text{(in conventional General Relativity)}$$

As a result the notions of space-time and time intervals (which depend on the geometry through General Relativity) become uncertain. Thus, according to Wigner, and others, the concept of time intervals and space-time points becomes questionable.

The Two-Tier version of Quantum Gravity offers a way out of this dilemma. The gravitational force becomes stronger as one goes to shorter distances (higher energies) down to a distance (up to an energy) whose scale is set by M_c. At shorter distances (higher energies) the gravitational force becomes weaker and declines to zero at zero distance. Thus at very high energy the gravitational field fluctuations (Δh) are at worst inversely proportional to the energy (and probably decline by a higher power of

[59] E. P. Wigner, Rev. Mod. Phys. **29**, 255 (1957); J. Math. Phys. **2**, 207 (1961).

inverse energy.) (The same considerations would apply if one chooses to consider fluctuations in the Riemann-Christoffel symbols.)

$$\Delta h < c_1/E < c_1/(\Delta E) \quad \text{(in Two-Tier Quantum Gravity)} \qquad (40.25)$$

where c_1 is a constant. Thus Wigner's conclusion does not hold in the Two-Tier version of Quantum Gravity as gravitational fluctuations actually become smaller at energies above a critical energy whose scale is set by M_c.

In fact, combining the above equations we see

$$c_1\Delta t/\Delta h \geq \hbar$$

at sufficiently high energy. Therefore the time uncertainty Δt, and the gravitational field fluctuations Δh, can both decrease while maintaining the energy-time uncertainty relation. *Thus the notion of a space-time point "is saved" in Two-Tier quantum gravity.*

40.11 Vacuum Fluctuations in the Gravitation Fields

While the expectation value of the free graviton field $h_{\mu\nu\text{conv}}(x)$ (weak field approximation) is zero in a conventional quantum field theoretic approach:

$$<0|h_{\mu\nu\text{conv}}(x)|0> = 0 \qquad (40.26)$$

the vacuum fluctuations of the *conventional* quantum graviton field is quadratically divergent since

$$<0|h_{\mu\nu\text{conv}}(x)h_{\alpha\beta\text{conv}}(x)|0> = \int d^3p\, b'_{\mu\nu\alpha\beta}(p)/[(2\pi)^3\, 2\omega_p] = \infty \qquad (40.27)$$

where $b'_{\mu\nu\alpha\beta}(p)$ is a rational function of the momentum p.

In "Two-Tier" quantum field theory we find

$$<0|h_{\mu\nu}(X)h_{\alpha\beta}(X)|0> = \int d^3p\, b'_{\mu\nu\alpha\beta}(p)\, e^{-p^ip^j\Delta_{Tij}(0)}/[(2\pi)^3 2\omega_p] = 0 \qquad (40.28)$$

since the exponential factor in the integrand is $-\infty$. The exponent contains

$$\Delta_{Tij}(z) = \int d^3k\, e^{-ik.z}(\delta_{ij} - k_ik_j/\mathbf{k}^2)/[(2\pi)^3 2\omega_k]$$

Thus the vacuum fluctuations of $h_{\mu\nu}$ are zero in "Two-Tier" quantum field theory and, correspondingly, the weak field Two-Tier quantization of Quantum Gravity is consistently finite (and weak in perturbation theory calculations.)

40.12 Two-Tier Features in D-Dimensional Space-Time (such as the Megaverse)

Since a field, quantized in D-dimensional conventional coordinates (D > 4), would lead to divergences in perturbation theory calculations, we can use D-dimensional Two-Tier coordinates to avoid divergences in perturbation theory:

$$Y^i(y) = y^i + i\, Y_u{}^i(y)/M_u{}^{D/2} \tag{40.29}$$

where $Y_u{}^i(y)$ for i = 1, …, D is a D-dimensional free gauge field and M_u is a mass of the order of the Planck mass or greater. The $Y_u{}^i(y)$ term adds a quantum field to the D coordinates making them a set of quantum coordinates. Quantum coordinate derivatives are defined by

$$\partial_i = \partial/\partial Y^i(y) = \partial/\partial(y^i - Y_u{}^i(y)/M_u{}^{D/2}) \tag{40.30}$$

The use of these coordinates to quantize particle fields leads to a completely finite perturbation theory. We applied them in Blaha (2017b) to create a finite fundamental theory of mater. We applied them to fields in the Megaverse[60] to achieve a finite theory of Megaverse dynamics for elementary particles and universe particles.

The second quantization of a vector gauge field $V^i(y)$ is analogous to the second quantization of the electromagnetic field. The lagrangian density terms for the free $V^i(Y(y))$ fields is

$$\mathscr{L}_{Vu} = -\tfrac{1}{4} F_{Vu}{}^{ij}(Y(y))F_{Vuij}(Y(y)) \tag{40.31}$$

The lagrangian is

$$L_{Vu} = \int d^Dy\, \mathscr{L}_{Vu}(Y(y))$$

with

$$F_{Vuij} = \partial V_i(Y(y))/\partial Y^j(y) - \partial V_j(Y(y))/\partial Y^i(y)$$

where the values of i and j range from 1 to D in this section.

The equal time commutation relations, using the Dth coordinate as the time coordinate, are specified in the usual way:

$$[V^i(Y(\mathbf{y}, y^0)), V^j(Y(\mathbf{y}', y^0))] = [\pi^i(Y(\mathbf{y}, y^0)), \pi^j(Y(\mathbf{y}', y^0))] = 0$$
$$[\pi_j(Y(\mathbf{y}, y^0)), V_k(Y(\mathbf{y}', y^0))] = -i\,\delta^{(D-1)tr}{}_{jk}(Y(\mathbf{y},0) - Y(\mathbf{y}',0))$$

where

$$\pi_u{}^k = \partial\mathscr{L}_{Vu}(V(Y(y)))/\partial V_k{}'(Y(y))$$
$$\pi_u{}^D = 0$$

for k = 1, … , (D – 1), and

$$\delta^{(D-1)tr}{}_{jk}(\mathbf{y} - \mathbf{y}') = \int d^{(D-1)}k\, e^{i\,\mathbf{k}\cdot(\mathbf{Y}(y,0) - \mathbf{Y}(y',0))}(\delta_{jk} - k_jk_k/\mathbf{k}^2)/(2\pi)^{D-1} \tag{40.32}$$
$$V_k{}'(Y(y)) = \partial V_k(Y(y))/\partial y^{1D}$$

for j, k = 1, 2, … , (D – 1).

---

[60] Blaha (2017c).

If we choose the Radiation gauge for $V_K(Y(y))$:

$$V^D(Y(y)) = 0$$
$$\partial V^j(Y(y))/\partial Y^j(y) = 0 \qquad (40.33)$$

for j = 1, 2, … , (D – 1) then (D – 2) degrees of freedom (polarizations) are present in the vector potential.[61] The Fourier expansion of the vector potential $V^i(Y(y))$ is:

$$V^i(Y(y)) = \int d^{(D-1)}k\, N_{0V}(k) \sum_{\lambda=1}^{D-2} \varepsilon^i(k, \lambda)[a_V(k,\lambda) :e^{-ik\cdot Y(y)}: + a_V^\dagger(k,\lambda) :e^{ik\cdot Y(y)}:] \quad (40.34)$$

for i = 1, … , (D – 2) where

$$N_{0V}(k) = [(2\pi)^{(D-1)} 2\omega_k]^{-\frac{1}{2}}$$

and (since the field is massless)

$$k^D = \omega_k = (\mathbf{k}^2)^{\frac{1}{2}}$$

where k^D is the energy, and where the $\varepsilon^i(k, \lambda)$ are the polarization unit vectors for λ = 1, … , (D – 2) and $k^\mu k_\mu = k^{D\,2} - \mathbf{k}^2 = 0$.

The commutation relations of the Fourier coefficient operators are:

$$[a_V(k,\lambda), a_V^\dagger(k',\lambda')] = \delta_{\lambda\lambda'}\delta^{D-1}(\mathbf{k} - \mathbf{k}')$$
$$[a_V^\dagger(k,\lambda), a_V^\dagger(k',\lambda')] = [a_V(k,\lambda), a_V(k',\lambda')] = 0$$

and the polarization vectors satisfy

$$\sum_{\lambda=1}^{D-2} \varepsilon_i(k, \lambda)\varepsilon_j(k, \lambda) = (\delta_{ij} - k_i k_j/\mathbf{k}^2)$$

The V^μ Feynman propagator is

$$iD_F^{trTT}(y_1 - y_2)_{jk} = \langle 0|T(V_j(Y(y_1))V_k(Y(y_2)))|0\rangle \qquad (40.35)$$

$$= -ig_{jk} \int \frac{d^Dk\, e^{-ik.(y_1 - y_2)}\, R(\mathbf{k}, y_1 - y_2)}{(2\pi)^D\, (k^2 + i\varepsilon)}$$

where g_{jk} is the D-dimensional Lorentz metric and where $R(\mathbf{k}, y_1 - y_2)$ is given by

$$R(\mathbf{k}, y_1 - y_2) = \exp[-k^i k^j \Delta_{Tij}(y_1 - y_2)/M_u^D]$$
$$= \exp\{-k^2[A(v) + B(v)\cos^2\theta] / [(2\pi)^{D-2} M_u^4 z^2]\}$$

[61] Note we use the Radiation gauge for $Y^\mu(y)$ also.

where k^2 is *the sum of the squares of the D – 1 spatial components* with

$$z^{\mu} = y_1^{\mu} - y_2^{\mu}$$
$$z = |\mathbf{z}| = |\mathbf{y_1} - \mathbf{y_2}|$$
$$k = |\mathbf{k}|$$
$$v = |z^0|/z$$
$$A(v) = (1 - v^2)^{-1} + .5v \ln[(v - 1)/(v + 1)]$$
$$B(v) = v^2(1 - v^2)^{-1} - 1.5v \ln[(v - 1)/(v + 1)]$$
$$\mathbf{k} \cdot \mathbf{z} = kz \cos\theta$$

and $|\mathbf{k}|$ denoting the length of a spatial (D – 1)-vector **k** while $|z^0|$ is the absolute value of $z^0 \equiv z^D$.

As the above equations indicate, the Gaussian damping factor R(k, z) for *all* large spatial momentum k^j is the same for both the positive and negative frequency parts of the (Two Tier) V Feynman propagator. We are assuming the spatial momentum is real-valued in this discussion. It is also important to note that R(k, z) does not depend on $k^0 = k^D$ (in the V and Y_u Radiation gauges) and thus the integration over k^0 proceeds in the usual way to produce time-ordered positive and negative frequency parts.

The Gaussian exponential factor in *all* spatial coordinates causes the Feynman propagator to be finite and, together with the Gaussian factor in universe particle propagators, causes all perturbation theory calculations when interactions are introduced to be finite as we have seen in Blaha (2017b).

For small momentum much less than M_u then $R(\mathbf{k}, y_1 - y_2) \rightarrow 1$ and the Feynman propagator is the "normal" propagator of conventional D-dimensional quantum field theory. For large momentum the corresponding potential approaches r^{D-3} in contrast to the electromagnetic Coulomb potential r^{-1}. The V potential is highly non-singular at large energies.

Thus using Two-Tier Quantum Field Theory we can perform perturbation theory calculations that always yield a finite result.[62] This is not true if conventional Quantum Field is used.[63]

41. PseudoQuantum Field Theory

PseudoQuantum Field Theory (and its Quantum Mechanics analogue CQ Mechanics[64]) originates in the need to second quantize in unusual coordinate systems,

[62] In particular, the fermion triangle divergence (anomaly) does not occur in our Two Tier Quantum Field Theory of the fermion sector. Thus there is no requirement for axion-like particles in the Megaverse (or in universes) although the possible existence of this type of particle is not ruled out.

[63] Blaha (2005a) provides a complete discussion of Two-Tier Quantum Field Theory.

[64] See Blaha (2016f) for CQ Mechanics, which encompasses both classical mechanics and quantum mechanics, and provides a method of rotating between them. It has applications to transitions between Quantum/Semi-Classical Entanglement, and Quantum/Classical Path Integrals, and Quantum/Classical Chaos.

and in curved space-time coordinate systems. The paper in Appendix 1-A provide a detailed introduction to PseudoQuantum Field Theory.

In this subsection we point out its advantages in a variety of field theory contexts that are relevant for the Unified SuperStandard Theory. The advantages of PseudoQuantum Field Theory are:

1. Quantization in any coordinate system in flat or curved space-times with an invariant definition of asymptotic particle states. An n particle asymptotic state in one coordinate system is a unitarily equivalent n particle asymptotic state in any other coordinate system. Therefore particle number is invariant under change of coordinate system. This is important for the Unified SuperStandard Theory in curved space-times. It is also important for quantization in higher dimensional Euclidean spaces such as the Megaverse. The method was developed in the late 1970's by the author to provide a quantization procedure which supports a unique particle interpretation of states in arbitrary non-static space-times where no global timelike coordinate (Killing vector) exists. PseudoQuantum Field Theory which we developed in a series of books[65] also can be formulated in the Megaverse. Thus we can use it in the Megaverse to implement the Higgs Mechanism to generate particle masses and symmetry breaking.

2. PseudoQuantum Field Theory enables one to define Higgs particle dynamics in such a way that a non-zero vacuum expectation value cleanly separates from the quantum field part of the Higgs fields. This technique can be used in symmetry breaking mechanisms, mass generation, and possible generation of coupling constants as vacuum expectation values.

3. It supports the canonical definition of higher derivative field theories through the use of the Ostrogradski bootstrap. See Appendix B where a fourth order theory of the Strong interaction is defined that has color confinement and a linear r potential. The potential part of this theory was used by the Cornell group to calculate the Charmonium spectrum. (See Blaha (2017b) for details.)

An associated advantage of using PseudoQuantum Field Theory is that it provides for retarded propagators and an Arrow of Time.

41.1 General Case of PseudoQuantization in Differing Coordinate Systems

Papers in Appendix A describe the PseudoQuantization procedure that relates second quantizations in differing coordinate systems. We can epitomize the general concept in the following short example.

---

[65] See Blaha (2017b) for the discussion of the PseudoQuantum field theory formalism for Higgs particles in our Extended Standard Model. See chapter 20 of Blaha (2017b), and earlier books, for a more detailed view than that presented here.

Consider the case of a scalar particle in D space-time dimensions that we second quantize in coordinate system denoted 1 with coordinates x based on a timelike Killing vector

$$\varphi(x) = \sum_{\alpha} [\chi_\alpha(x)A_\alpha + \chi_\alpha{}^*(x)A_\alpha{}^\dagger] \quad (41.1)$$

where the $\chi_\alpha(x)$ are positive frequency with respect to a definition of positive frequency within a universe – following the notation of Appendix A.

Consider now the second quantization of the particle field in a second coordinate system denoted 2 with coordinates y based on a different timelike Megaverse Killing vector

$$\varphi(y) = \sum_{\beta} [\psi_\beta(y)b_\beta + \psi_\beta{}^*(y)b_\beta{}^\dagger] \quad (41.2)$$

where the $\psi_\beta(y)$ are positive frequency with respect to 2's definition of positive frequency.

Comparing above definitions we see the difference in the definition of the coordinates used in the field expansions as well as the implicit difference in the definitions of positive frequency. To relate the quantizations to each other, we must use the relation between the x and y coordinates:

$$y_i = f_i(x)$$

or, in vector form,

$$y = f(x)$$

for i = 1, 2, … , D. Thus

$$\varphi(f(x)) = \sum_{\beta} [\psi_\beta(f(x))b_\beta + \psi_\beta{}^*(f(x))b_\beta{}^\dagger] \quad (41.3)$$

Inverting the above equations to obtain the relation of the fourier coefficient operators we see:

$$A_\alpha = \sum_{\beta} [C_{\alpha\beta}\, b_\beta + C'_{\alpha\beta}\, b_\beta{}^\dagger]$$

where $C_{\alpha\beta}$ and $C'_{\alpha\beta}$ are c-number functions of α and β:

$$C_{\alpha\beta} = (\chi_\alpha(x), \varphi(f(x)) \quad (41.4)$$
$$C'_{\alpha\beta} = (\chi_\alpha{}^*(x), \varphi(f(x))$$

The above equations imply an N particle state in one coordinate system will appear as a superposition of states of various numbers of particles in the other coordinate system IF the standard quantum field theory formulation is used.

TO REMEDY this situation – which we take to be unphysical – we must reformulate quantum field theory using the PseudoQuantum formulation presented Appendix 1-A. The scalar particle case is discussed in Appendix 1-A.

The conclusions of that section, and the sections following it in Appendix 1-A, are:

1. One can define corresponding unitarily equivalent particle states in two quantizations with invariant particle numbers.

2. The fourier coefficient operators of the two quantizations are related by Bogoliubov transformations and are unitarily equivalent.

3. The group of the local Bogoliubov transformations is an infinite tensor product of $SU_{1,1}$ groups.
4. The vacuums of the particles are invariant under Bogoliubov transformations that relate the Megaverse and the universe quantizations.

5. Unitarily equivalent perturbation theories of both quantizations can be defined.

We now consider the case of Two-Tier PseudoQuantization, and then turn to various applications of PseudoQuantization.

41.2 Two-Tier PseudoQuantum Field Theory

The combination of the Two-Tier procedure with the PseudoQuantization procedure leads to a somewhat more complicated situation. In principle, both are required for a Unified SuperStandard Theory in any coordinate system in flat or curved space-times in any number of dimensions. However their direct combination is both complicated and unphysical.

The main purpose of PseudoQuantization is to have particle number invariance under a change of coordinate system. Two-Tier Field Theory 'cloaks' each particle in infinite 'clouds' of Y^μ quanta as Fig. 40.1 illustrates. We define PseudoQuantization as implementing particle number invariance for 'bare' particles without their clouds of Y^μ quanta. Thus an asymptotic particle state of n particles (neglecting its Y^μ quanta cloud) remains a unitarily equivalent n particle state (neglecting its Y^μ quanta cloud) under a change of coordinate system.

To implement this concept we first define quantizations of a particle in coordinate systems without Two-tier quanta. We then 'dress' the quantizations by replacing the coordinates y^μ in each coordinate system with the corresponding Two-Tier coordinates:

$$y^\mu \rightarrow X^\mu(y) = y^\mu + i\, Y^\mu(y)/M_c^2 \qquad (41.5)$$

It appears the most convenient gauge in each coordinate system is the Lorentz gauge:

$$\partial Y^{\mu}/\partial y^{\mu} = 0 \tag{41.6}$$

We now briefly consider the case of a scalar particle PseudoQuantization. This case is considered in more detail in Appendix 1-A. Following Appendix 1-A we must introduce two fields $\varphi_1(y)$ and $\varphi_2(y)$ with the free fields' lagrangian

$$\mathscr{L}(y) = \partial^{\mu}\varphi_1\partial_{\mu}\varphi_2 - \tfrac{1}{2}\,\partial^{\mu}\varphi_1\partial_{\mu}\varphi_1 - m^2\varphi_1\varphi_2 + \tfrac{1}{2}\,m^2\,\varphi_1^{\,2} \tag{41.7}$$

in a coordinate system with coordinates y. Then following the steps indicated in Appendix 1-A from eq. 7 onward we arrive at a PseudoQuantum formulation in the coordinate system with coordinates y that is unitarily equivalent to that of a different coordinate system defined a similar manner.

From eq. 43 onwards we can replace the c-number coordinates x and y with Two-Tier coordinates of the form

$$X^{\mu}(y) = y^{\mu} + i\,Y^{\mu}(y)/M_c^{\,2}$$

and proceed to calculate propagators and perturbation theory diagrams.

Thus we have a straight-forward procedure to unite the PseudoQuantum formalism with Two-Tier coordinates to obtain finite perturbation theory results with unitary equivalence to quantization in other coordinate systems in both flat and curved space-times.

The use of two fields per particle of PseudoQuantum field theory will be seen to part of the applications consider in the remainder of this subsection. We will put aside the consideration of quantizations in other coordinate systems in what follows to keep the presentation as simple as possible.

41.3 PseudoQuantum Higgs Scalar Particle Field Theory in D-dimensional Space-Time

41.3.1 The Enigma of Higgs Particles and the Higgs Mechanism

In our previous work on the Standard Model, and its generalization to The Unified SuperStandard Theory described in a series of books entitled *Physics is Logic* ..., we showed that the fermion spectrum results from Complex Special Relativity, the gauge interactions result from the Reality group, the fermion generations result from the Generation group, and the Theory of Everything results from a combination with Complex General Relativity. The Higgs particles and the Higgs Mechanism were inserted to generate particle masses and symmetry breaking effects.

Whence comes Higgs particles? A more fundamental cause has not been suggested until our analysis, which is presented here. So the Higgs sector appeared to be an expedient mechanism to insert much needed symmetry breaking and masses into the theory.

There are a number of peculiarities in the implementation of the Higgs Mechanism:

1. First, it is selective in the sense that some gauge fields have associated Higgs particles and utilize the Higgs Mechanism, and some gauge fields do not have associated Higgs particles. In particular, the ElectroWeak gauge fields, the Generation group gauge fields, the Layer group fields, and the complex gravity Species gauge fields have associated Higgs particles. The strong interaction (gluon) gauge fields do not.

2. The Higgs potentials have a quadratic mass term of the "wrong" sign plus a quartic interaction term, which together, generate non-zero vacuum expectation values. They obviously accomplish their goal. But the source of these potentials, and why they have the same form, is unknown. One expects a fundamental principle should be operative here.

3. One can imagine creating a Higgs microscope at some super-accelerator. Using this microscope in the presence of a (classical) condensate could enable the Uncertainty Principle to be violated. This possibility, in the case of a microscope using electromagnetic fields, was the source of a heuristic argument for the need to quantize the electromagnetic field.[66]

4. The formulation of the Higgs Mechanism uses classical fields under the assumption that a path integral formulation justifies their use. While this may be true, the path integral formulation relies on implicit, unstated boundary conditions that obscure the physics of the quantum field theoretic nature of the mechanism. A direct quantum field theoretic study of the Higgs Mechanism is needed and would further elucidate its character.

5. Scalar fields have a cloud hanging over them that spin ½ fields do not. A spin ½ particle cannot transition to negative energy because there is a filled sea of negative energy particles. No additional particles can fall into the sea due to the Pauli Exclusion Principle that forbids two fermions with the same 4-momentum and quantum numbers. In the case of scalar particles the Pauli Exclusion Principle does not apply and so a *filled* negative energy sea of scalar particles is not possible and positive energy scalar particles can transition to negative energy without hindrance. This problem has been "resolved" by an appropriate definition of the scalar particle vacuum to exclude transitions to negative energy. But the rationale for the definition is lacking. Dirac was asked about this issue many years ago. He said he had a solution to the problem. However he did not present it – in keeping with his well-known taciturn nature. So the issue remains an open question.

---

[66] Heitler (1954) p. 86 provides a good discussion of the need to quantize the electromagnetic field.

For the above reasons we will show that a more satisfactory method of achieving the goals of mass generation and symmetry breaking exists.[67] This method relies on a larger Fock space that enables the appearance of a vacuum expectation value for Higgs particles to be understood within a truly quantum framework. More importantly, this method is a consequence the PseudoQuantization procedure described above that enables unitarily equivalent quantizations in different coordinate systems. So a profound fundamental justification for our Higgs boson formulation exists. One major consequence of this approach is the appearance of a local Arrow of Time – a concept that has been a subject of interest for over one hundred years. Another consequence is a rationale for ElectroWeak Higgs bosons and for their absence for the strong (gluon) interaction.

41.3.2 PseudoQuantization of Scalar Particles

We now consider the PseudoQuantization[68] of a scalar particle field that will become a Higgs particle with a non-zero vacuum expectation value.[69] We begin by defining two fields that correspond to the scalar particle: $\varphi_1(x)$ and $\varphi_2(x)$.[70] These fields will be assumed to have the equal time commutators

$$
\begin{aligned}
&[\varphi_i(x), \pi_j(y)] = i(1 - \delta_{ij})\delta^3(\mathbf{x} - \mathbf{y}) \\
&[\varphi_i(x), \varphi_j(y)] = 0 \\
&[\pi_i(x), \pi_j(y)] = 0
\end{aligned}
\tag{41.8}
$$

where δ_{ij} is the Kronecker δ and where $\pi_i(x)$ is the canonically conjugate momentum to $\varphi_i(x)$. The fields $\varphi_1(x)$ and $\pi_1(y)$ will be observable classical fields. The fields $\varphi_2(x)$ and $\pi_2(y)$ will not be observables so that $\varphi_1(x)$ and $\pi_1(y)$ can both be sharp on the set of physical states.

We now specify the lagrangian density for a scalar Klein-Gordon particle:

$$\mathcal{L} = \partial\varphi_1/\partial x_\mu \partial\varphi_2/\partial x^\mu \tag{41.9}$$

with hamiltonian density

$$\mathcal{H} = \pi_1\,\pi_2 + \partial\varphi_1/\partial x_i \partial\varphi_2/\partial x^i$$

---

[67] In the Extended Standard Model of Blaha (2015a) we have shown that the basic particles have a mass, the Landauer mass, so that the theory is symmetry violating from the very start. We have also shown that our Two-Tier formalism for quantum field theories always yields finite results in perturbation theory calculations – making the renormalization approach of t'Hooft and others, which relied on initially massless gauge fields, unnecessary.

[68] PseudoQuantization in a D-dimensional space-time is described in Blaha (2017c). This discussion is relevant to PseudoQuantization in the Megaverse, or in other universes.

[69] Much of this section appears in Blaha (2016c), and earlier books, as well as in S. Blaha, Phys. Rev. **D17**, 994 (1978). The case of fermion PseudoQuantization is also discussed in Appendix A – S. Blaha, Il Nuovo Cimento **49A**, 35 (1979).

[70] The subscripts on the fields are not gauge symmetry indices but simply identifiers distinguishing the fields from each other.

where i labels spatial coordinates, and $\pi_1 = \partial\varphi_2/\partial t$ and $\pi_2 = \partial\varphi_1/\partial t$. The lagrangian $\mathcal{L}$ is without a potential or mass term.

The lagrangian and hamiltonian for a massive scalar particle in this formalism are

$$\mathcal{L} = \partial\varphi_1/\partial x_\mu \partial\varphi_2/\partial x^\mu - m^2 \varphi_1\varphi_2 \quad (41.10)$$

with hamiltonian density

$$\mathcal{H} = \pi_1 \pi_2 + \partial\varphi_1/\partial x_i \partial\varphi_2/\partial x^i + m^2 \varphi_1\varphi_2$$

The fields can be fourier expanded in terms of creation and annihilation operators:

$$\varphi_i(\mathbf{x}, t) = \int d^3k\, [a_i(k)f_k(x) + a_i^\dagger(k)f_k^*(x)] \quad (41.11)$$

for i = 1, 2 where

$$f_k(x) = e^{-ik\bullet x}/(2\omega_k(2\pi)^3)^{1/2}$$

with $\omega_k = |\mathbf{k}|$.

The creation and annihilation operators satisfy the commutation relations:

$$[a_i(k), a_j^\dagger(k')] = (1 - \delta_{ij})\delta^3(\mathbf{k} - \mathbf{k'})$$
$$[a_i(k), a_j(k')] = 0$$
$$[a_i^\dagger(k), a_j^\dagger(k')] = 0$$

for i, j = 1, 2.

In this formulation the defining properties of a physical state are:

$$\varphi_1(x)|\Phi, \Pi\rangle = \Phi(x)|\Phi, \Pi\rangle \quad (41.12)$$
$$\pi_1(x)|\Phi, \Pi\rangle = \Pi(x)|\Phi, \Pi\rangle$$

where $\Phi(x)$ and $\Pi(x)$ are sharp on the states and thus classical fields with

$$\Phi(\mathbf{x}, t) = \int d^3k\, [\alpha(k)f_k(x) + \alpha^*(k)f_k^*(x)] \quad (41.13)$$

and correspondingly for $\Pi(x)$.

41.3.3 Vacuum States for Scalar (Higgs) Particles with Non-Zero Vacuum Expectation Values

When we implement the mass mechanism, Φ is constant. We can define a set of states

$$a_1(k)|\alpha\rangle = \alpha(k)|\alpha\rangle$$
$$a_1^\dagger(k)|\alpha\rangle = \alpha^*(k)|\alpha\rangle$$

and correspondingly a set of coherent states

$$|\alpha\rangle = C\exp\{\int d^3k\, [\alpha(k)a_2^\dagger(k) + \alpha^*(k)a_2(k)]\}|0\rangle \quad (41.14)$$

where C is a normalization constant and where the vacuum state $|0\rangle$ satisfies

$$a_1(k)|0> = a_1^\dagger(k)|0> = 0 \qquad (41.15)$$

$$a_2(k)|0> \neq 0 \qquad\qquad a_2^\dagger(k)|0> \neq 0$$

The dual vacuum state satisfies

$$<0|a_2(k) = <0|a_2^\dagger(k) = 0$$

$$<0|a_1(k) \neq 0 \qquad\qquad <0|a_1^\dagger(k) \neq 0$$

With this coherent state formalism, which gives purely classical fields and yet also has quantum fields through the use of φ_2 and its creation and annihilation operators, we now have the machinery to define a mass mechanism without the introduction of a potential whose origin can only be described as dubious.

For we can define a coherent state for some k as

$$|\Phi, \Pi> = C\exp\{[(2\pi)^3\omega_k/2]^{1/2}\Phi[a_2^\dagger(k) + a_2(k)]\}|0> \qquad (41.16)$$

where C is a normalization constant, which yields a non-zero vacuum expectation value:

$$\varphi_1(x)|\Phi, \Pi> = \Phi|\ \Phi, \Pi> \qquad (41.17)$$

where Φ is a constant. Evaluating a fermion interaction term we find a mass term emerges[71]

$$\psi\ (\varphi_1 + \varphi_2)\psi \ \rightarrow \bar{\psi}(\Phi + \varphi_2)\psi \qquad (41.18)$$

It generates a mass for an interaction with a gauge field of the form

$$A^\mu(\varphi_1 + \varphi_2)^2 A_\mu \ \rightarrow A^\mu(\Phi + \varphi_2)^2 A_\mu \qquad (41.19)$$

It also yields a quantum field theoretic interaction that would result in the production of ElectroWeak particles from these scalar fields. The production of Higgs particles that decay into ElectroWeak gauge particles has recently been found at CERN.

The present formalism provides a clean way to separate the vacuum expectation value of a scalar particle from its quantum field part in contrast to the Higgs Mechanism where one has to separate a Higgs field into parts manually.

41.3.4 Interpretation of Negative Energy Scalar Particle States

As we noted earlier, scalar particle physics has the problem of no barrier to the decay of positive energy states to negative energy states due to the absence of a Pauli Exclusion Principle for bosons. The PseudoQuantization procedure that we developed

---

71 When matrix elements with a "vacuum state" are taken.

in 1978 and describe here allows negative energy states as one would physically expect and raises the possibility of disastrous particle decays to negative energy. The above equations show that negative energy states are possible in this theory.

However they also show that combined positive and negative energy boson states can be interpreted as classical field states. In addition, the ability of any number of boson particles to have the same 4-momentum and quantum numbers shows that a *macroscopic* classical scalar field state can be constructed.

Thus we can view states containing negative energy particles as classical field states and thus solve[72] *the issue of interpreting negative energy particle states – a more satisfactory approach than the standard quantization procedure does – with due respect to Professor Dirac.*

We note that macroscopic many particle fermion states can only have one particle in any mode unlike bosons. Therefore we cannot use this formalism to create macroscopic classical fermion field states.[73] And the filled Dirac sea of negative energy fermions precludes the transition of a positive energy Dirac fermion to a negative energy state. *Thus there is a certain complementarity between fermions that cannot become classical fields but have a filled sea precluding decays to negative energy states, and bosons that can become classical fields but support decays to negative energy states.*

41.3.5 Contrast with Conventional Second Quantization of Scalar Particles

The PseudoQuantization procedure followed here uses different boundary conditions than the usual scalar particle quantization procedure. The essence of the difference is embodied in a comparison of the definition of the vacuum above and the definition of the conventional second quantized field vacuum:

$$a|0\rangle = 0 \qquad \text{Conventional Approach}$$
$$a^{\dagger}|0\rangle \neq 0$$

In the conventional approach the creation of negative energy boson states is eliminated *ab initio* whereas in our approach it is allowed in order to support classical field states with non-zero vacuum expectation values that are a form of classical field. While one cannot discredit the conventional choice for conventional scalar fields, one can see that our approach yields a physically more important result – particularly for Higgs fields – because it leads to an Arrow of Time *locally* – an important feature of physical phenomena that has been a subject of much discussion and dispute. One can say that the conventional approach sweeps the issue "under the rug" rather than seeking a deeper justification – differing from Dirac's implied notion that the issue merited attention. We will discuss the "Arrow of Time" within the framework of our PseudoQuantization approach later.

---

[72] Also a boson that has no interactions cannot transition from to a positive energy state to a negative energy state due to conservation of energy.

[73] However we can create PseudoQuantum fermion states. See S. Blaha, Phys. Rev. **D17**, 994 (1978) (reproduced in Appendix I) and references therein to earlier papers by the author.

41.3.6 Why Inertial Reference Frames are Special

The great physicists of the early 20th century raised numerous questions about Special Relativity after Einstein and Poincarè's discovery. Prominent among them was the question of why inertial reference frames are of especial importance in Special Relativity, and afterwards in General Relativity.

It appears that our formulation of the mass generation mechanism sheds significant light on the reason for the special prominence of inertial frames. Earlier we considered the case of a massless PseudoQuantized scalar. We now consider massive scalars since experiments at CERN have apparently discovered a Higgs particle with a 125 GeV/c mass. The above equations describe a massive scalar particle. If the scalar is massive, then the "vacuum" state that yields a non-zero expectation value must change to

$$|\Phi, \Pi\rangle = C\exp\{(2\pi)^3 m/2]^{1/2}[a_2^{\dagger}(\mathbf{0},m) + a_2(\mathbf{0},m)]\}|0\rangle \qquad (41.20)$$

to have operators for a particle of mass m in its rest frame. Then, having established this preferred frame for a Higgs particle, in The Unified SuperStandard Theory, and requiring that invariant intervals

$$ds^2 = dt^2 - d\mathbf{x}^2 \quad \text{(in rectangular coordinates)}$$

are unchanged by a (complex or real) Lorentz transformation, we find that inertial reference frames are singled out as "special" in the sense that they are the only accessible reference frames that can be generated by a Lorentz boost/transformation from the Higgs particle rest frame. *The Higgs particle vacuum state singles out the class of inertial reference frames.*

Thus Higgs particles play a central role in establishing the basis of physical reality.

41.3.7 PseudoQuantization Reveals More Physical Consequences than the Higgs Mechanism of Scalar Particles

Earlier we pointed out that our PseudoQuantization theory of Higgs particles reveals more physical consequences than the conventional approach, which implements the Higgs Mechanism by simply using a potential term that has a minimum at a non-zero vacuum expectation value. This section shows the major results of a properly implemented mechanism. We find a better explanation of the negative energy state problem of boson field theories. We find a local arrow of time that explains the direction of time that we, and all of nature, experiences. We find the reason why inertial reference frames have a special physical significance – a result long sought by physicists.

In addition we will see in chapter 11 that real gauge fields should have an associated Higgs particle, while necessarily complex gauge fields (the Strong interaction gauge field in The Unified SuperStandard Theory) do not have an associated gauge field. These results correspond to experimental reality.

41.3.8 The T Invariance Issues of Our PseudoQuantized Scalar Particle Theory

The PseudoQuantized scalar particle hamiltonian equations are invariant under time reversal $t \rightarrow t' = -t$. The 'new' vacuum states defined above break the time reversal invariance of the theory resulting in retarded particle propagators.

The hamiltonian equations

$$[H, \varphi_1(\mathbf{x}, t)] = -i\partial\varphi_1/\partial t \qquad (41.21)$$
$$[H, \varphi_2(\mathbf{x}, t)] = -i\partial\varphi_2/\partial t$$

are invariant under time reversal. If we define a time reversal operator transformation U then the time reversed equations are

$$[UHU^{-1}, \varphi_1(\mathbf{x}, -t)] = +i\partial\varphi_1(\mathbf{x}, -t)/\partial(-t)$$
$$[UHU^{-1}, \varphi_2(\mathbf{x}, -t)] = +i\partial\varphi_2(\mathbf{x}, -t)/\partial(-t)$$

The operator U, which is unitary, transforms H into **–H**. This operation is legal because the hamiltonian – in this case – is not positive definite and admits negative energy states.[74] Thus

$$[H, \varphi_1(\mathbf{x}, -t)] = -i\partial\varphi_1(\mathbf{x}, -t)/\partial(-t)$$
$$[H, \varphi_2(\mathbf{x}, -t)] = -i\partial\varphi_2(\mathbf{x}, -t)/\partial(-t)$$

and the time reversal invariance of the equations of motion is established for this case.

Time reversal invariance is broken by our choice of vacuum states. This choice is necessary to obtain classical field states as we showed earlier. A demonstration of the time reversal symmetry breaking is presented later where we show theory has retarded propagators for particle propagation to and from asymptotic states.

Within the interaction region the particle propagators are the sum of retarded and advanced parts that combine to yield principle value propagators – not Feynman propagators. Many years ago Feynman and Wheeler championed principle value propagators for electrodynamics to obtain an action-at-a distance theory of Quantum Electrodynamics. While their theory, and ours, differ from the standard quantum field theory approach there is no reason to view them as faulty, or having serious physical defects. The only question is whether nature chooses conventional quantum field theory or PseudoQuantized quantum field theory. In our case the need for a classical scalar particle non-zero vacuum expectation value strongly motivates our choice of pseudoquantized Higgs particles.

41.3.9 Retarded Propagators for Our Quantized Higgs Particles

In the previous section we pointed out that our PseudoQuantization Higgs theory has an arrow of time due to its boundary conditions as expressed by its definition of the vacuum state and its dual. In this section we will show that the theory uses retarded propagators for propagation to and from the interaction region to asymptotic in-states

---

[74] Unlike the usual case of second quantized Klein-Gordon quantum field theory.

and out-states. Within an interaction region the theory uses half-retarded – half-advanced propagators. We discuss aspects of the perturbation theory and propagators of our scalar particles in this chapter.

First we note that in-states at $t = -\infty$ are composed of superpositions of $a_2(k)$ and $a_2^\dagger(k)$ creation and annihilation operators:

$$a_2(k)|0> \neq 0 \qquad\qquad a_2^\dagger(k)|0> \neq 0$$

while the out-states composed of superpositions of $a_1(k)$ and $a_1^\dagger(k)$ creation and annihilation operators:

$$<0|a_1(k) \neq 0 \qquad\qquad <0|a_1^\dagger(k) \neq 0$$

Consequently when in-state particles (x_1) propagate into the interaction region (x_2) the relevant propagators are retarded propagators with the form

$$\begin{aligned} G_{in}(x_2, x_1) &= <0|T(\varphi_{1\,in}(x_2), \varphi_{2\,in}(x_1))|0> \\ &= \theta(x_{20} - x_{10})<0|[\varphi_{1\,in}(x_2), \varphi_{2\,in}(x_1)]\,|0> \end{aligned} \tag{41.22}$$

This is a manifestly retarded propagator. The choice of vacuums clearly results in a time asymmetry giving a retarded propagation reflecting the familiar Arrow of Time.

A similar situation prevails for propagation to out-states (x_3) from the interaction (x_2) region:

$$\begin{aligned} G_{out}(x_3, x_2) &= <0|T(\varphi_{1\,out}(x_3), \varphi_{2\,out}(x_2))|0> \\ &= \theta(x_{30} - x_{20})<0|[\varphi_{1\,out}(x_3), \varphi_{2\,out}(x_2)]\,|0> \end{aligned} \tag{41.23}$$

Within the interaction region the Higgs particles have principle value propagators.

Thus we find PseudoQuantized Higgs particles embody a local Arrow of Time. The locality of the Arrow of Time is embodied in all the particles that interact with the Higgs particle. Since the mass of *every* particle – bosons and fermions – has a Higgs contribution, and thus *every* particle interacts with the Higgs particles, the Arrow of Time permeates The Unified SuperStandard Theory as well as the more familiar Standard Model known from experiment.

41.3.10 The Local Arrow of Time

In the *Physics is Logic* series of monographs we saw that complex coordinates led to the form of the fermion spectrum, that the mapping of complex coordinates to real-valued coordinates yielded the Reality group and The Unified SuperStandard Theory gauge interactions, that Complex General Relativity led to Higgs particles that were directly united with elementary particle masses and gave us the equality of inertial mass and gravitational mass. Later we will see the reduction of complex gauge fields to real gauge fields explains the appearance of Higgs fields in The Unified SuperStandard Theory.

The PseudoQuantization procedure leads to retarded Higgs field propagators and thence to a *local* arrow of time. Many arguments have been put forward over the past hundred plus years for the Arrow of Time. Many arguments based on Statistical Mechanics, Entropy, and Boltzmann's statistical atomic theory have suggested the Arrow of Time is a global statistical consequence. This view seems to contradict the results of elementary particle experiments where a *local* Arrow of Time is evident.

Our rationale for the Arrow of Time begins with retarded Higgs fields. Then we note that Higgs field quantum interactions appear for all fermions and gauge particles. Thus all particle interactions are imbued with an Arrow of Time. Particles united to form macroscopic matter inherit their combined Arrows of Time producing the global Arrow of Time we experience.

Thus our PseudoQuantization approach offers a more satisfactory solution of the origin of the Arrow of Time.

It is remarkable that complex quantities – coordinates and fields – through the Higgs phenomena that we have considered, lead to the equality of inertial mass and gravitational mass, and an Arrow of Time. This unity of mass and time phenomena may reflect the deeper fact that we can have no practical Arrow of Time if all particles were massless, for particle dynamics at light speed would then be pointless. This view has been expressed by DeWitt, Unruh, and others who have pointed out that, physically, time is meaningful and measurable only if masses exist; the larger the mass, the more accurate the time measurement in principle.[75]

41.3.11 Space-Time Dependent Particle Masses

It is possible that the ultimate Unified SuperStandard Theory has masses that evolve with time and may also be spatially varying – different values in different parts of the universe. Presently there is no decisive evidence for this possibility although astrophysical studies continue. In this section we will describe the mechanism for space-time dependent masses.

Consider a classical field (time and spatially varying):

$$\Phi(\mathbf{x}, t) = \int d^3k\, [\alpha(k)f_k(x) + \alpha^*(k)f_k{}^*(x)] \qquad (41.24)$$

If we define the coherent vacuum state

$$|\alpha\rangle = C \exp\{\int d^3k\, [\alpha(k)a_2{}^\dagger(k) + \alpha^*(k)a_2(k)]\}|0\rangle \qquad (41.25)$$

then

$$\varphi_1(x)|\Phi, \Pi\rangle = \Phi(x)|\Phi, \Pi\rangle$$
$$\pi_1(x)|\Phi, \Pi\rangle = \Pi(x)|\Phi, \Pi\rangle$$

where

$$\varphi_i(\mathbf{x}, t) = \int d^3k\, [a_i(k)f_k(x) + a_i{}^\dagger(k)f_k{}^*(x)] \qquad (41.26)$$

[75] No mass, no clock; no clock, no physical time. See Blaha (2015a) pp. 368-371 for a discussion including comments by DeWitt and Unruh.

for i = 1, 2 and where

$$f_k(x) = e^{-ik\cdot x} /(2\omega_k(2\pi)^3)^{1/2}$$

with ω_k equal to the energy.

41.3.12 Inertial Mass Equals Gravitational Mass

From the days of Newton through Einstein[76] to the present the equality of gravitational mass and inertial mass has been a topic of interest. Mach, who played an important role, in this ongoing discussion, thought distant masses in the universe were the source of the equality. However the origin of the equality, which has been shown experimentally to very high accuracy, remained uncertain until the *Physics is Logic* series of books, in which we showed the interconnection of the Unified SuperStandard Theory and Complex Gravitation via Higgs generated masses that united gravitational and inertial mass.

In Blaha (2016h) we showed that a Complex General Relativity transformation can be factored into the product of a complex-valued transformation and a real-valued General Coordinate transformation. The set of complex valued transformations form a U(4) group that we called the General Coordinate Reality group. Later we will define the Internal Symmetry Species Group as the corresponding analogue. The Species Group has gauge fields that undergo spontaneous symmetry breaking and generate contributions to all fermion masses.

Since fermion field masses are now sums of ElectroWeak Higgs contributions, Generation group Higgs contributions, Layer group Higgs contributions, and Species group contributions, and since the gravitational Higgs fields appear in all fermion masses, the equality of inertial and gravitational mass is proven. The gravitational Higgs particles' equations depend, in part, on the gravitational field by Blaha (2016h) and so set the mass scale of gravitational mass, and thereby of all Higgs mass contributions. They set the scale of inertial masses equal to the scale of gravitational masses. ***Since an expression cannot mix mass scales, the gravitational mass scale must be the same as the inertial mass scale. Inertial Mass equals gravitational mass.***

We have established the equality of inertial and gravitational mass at the short distance quantum level. In our view, this explanation is far more satisfying than basing the equality on a combination of large distance phenomena and quantum phenomena. As Einstein and Weyl have pointed out, all fundamental physics phenomena should be based on a local theory. Complex Gravity as we have constructed it, combined with the Unified SuperStandard Theory, furnishes a completely local basic Theory of Everything.

---

[76] For example, Einstein and Grossman in 1913 stated, "The theory herein described originates in the conviction that the proportionality between the inertial and gravitational mass of a body is an exact law of nature that must be expressed as a foundation principle of theoretical physics."

The equation above contains a coherent state $|\alpha>$ for a time and spatially varying mass. The above equations can be generalized to the case of multiple space-time varying masses.[77]

$$|\Phi_1,\Phi_2, \ldots ,\Phi_n;\Pi_1,\Pi_2, \ldots ,\Pi_n> = C \prod_{i=1}^{n} \exp\{\int d^3k\, [\alpha_i(k)a_{2i}^\dagger(k) + \alpha_i^*(k)a_{2i}(k)]\}|0> \quad (41.27)$$

Then all n mass vacuum expectation values are space-time dependent:

$$\varphi_{1i}(x) \mid \Phi_1, \Phi_2, \ldots , \Phi_n; \Pi_1, \Pi_2, \ldots , \Pi_n> = \Phi_i(x) \mid \Phi_1, \Phi_2, \ldots , \Phi_n; \Pi_1, \Pi_2, \ldots , \Pi_n> \quad (41.28)$$

Thus our formalism can accommodate space-time varying masses should they be found in the Cosmos.

41.3.13 Benefits of the PseudoQuantization Method

In this book, and in earlier work, we showed that a more physically satisfactory method for avoiding the negative energy state problem exists. This method relies on the use of a larger Fock space in which negative energy states (or partially negative energy states) are interpreted as states containing classical fields or a mix of classical fields and individual boson particles. This approach resolves the negative energy boson issue and provides a common framework for boson particles and classical boson fields.

One consequence of the PseudoQuantization method is that it enables the appearance of a vacuum expectation value for Higgs particles (a constant classical field) to be understood within a truly quantum framework. Another major consequence of this approach is the appearance of a *local* Arrow of Time due to the Higgs mass generation mechanism – a concept that has been a subject of interest for over one hundred years. A macroscopic arrow of time is often described as a statistical result. But our approach yields an arrow of time at the single particle level.

The conventional approach to boson field quantization sweeps these issues "under the rug" rather than seeking a deeper justification. It differs from Dirac's implied notion that the issue merited attention.

Another important consequence of the PseudoQuantization method is that it singles out inertial reference frames when applied to the case of Higgs particles.

Yet another more subtle consequence of boson PseudoQuantization is that it provides a rationale/explanation for the presence of ElectroWeak Higgs bosons, *and for their absence for the strong (gluon) interactions. The question of why there are no strong interaction Higgs bosons has not been previously considered to the best of this author's knowledge.*

41.1 Two-Tier Formulation and PseudoQuantization in the Megaverse

These subjects are considered in Blaha (2020c) in chapter 68 as well as in earlier books by the author.

---

[77] The "vacuum" state $|0>$ also implicitly has factors for the vacuum expectation values used for fields that give masses to fermions and vector bosons as described in Blaha (2016h).

5. QUeST-UST Lagrangian Formulation based on the Riemann-Christoffel Curvature Tensor

This chapter describes the basic structure of the gauge vector boson dynamics and interactions. It is abstracted from chapter 53 of Blaha (2020c) and earlier books by the author. The most significant change is the insertion of U(1)⊗U(1) Fermion group terms: one for the NORMAL sector and one for the DARK sector.

Noteworthy features include:

1. Higher derivative Strong interactions giving a confining r potential for quarks.
2. Higher derivative Gravitation giving gravity modifications of the 1/r potential at various large distances; intra-galactic and inter-galactic. The modifications are reminiscent of MOND.
3. It is a PseudoQuantum formulation with two fields for each vector boson.

In view of our goal of defining a unified theory of elementary particles and General Relativity we begin by defining a Riemann-Christoffel curvature tensor which we will use to construct a lagrangian for the theory.

5.1 The Covariant Derivative

The covariant derivative[78] which appears in fermion and gravitation equations uses the vector boson 12-vector:

$$ {}^a\mathbf{A}_I{}^\mu(x) = ({}^ag_1{}^a\mathbf{A}_{SU(3)}{}^\mu(x_C),\ {}^ag_2{}^a\mathbf{W}^\mu(x)\ ,\ {}^ag_3{}^a\mathbf{A}_E{}^\mu(x),\ {}^ag_4{}^a\mathbf{W}_D{}^\mu(x),\ {}^ag_5{}^a\mathbf{A}_{DE}{}^\mu(x),\ {}^ag_6{}^a\mathbf{U}^\mu(x),\ {}^ag_7{}^a\mathbf{V}^\mu(x), \\ {}^ag_8{}^a\mathbf{A}_{DSU(3)}{}^\mu(x_C),\ {}^ag_9{}^a\mathbf{U}_D{}^\mu(x),\ {}^ag_{10}{}^a\mathbf{V}_D{}^\mu(x),\ {}^ag_{11}{}^a\mathbf{Y}^\mu(x),\ {}^ag_{12}{}^a\mathbf{Y}_D{}^\mu(x)\ ) \qquad (5.1) $$

where a labels the layer, a = 1, 2, 3, 4. We label the respective coupling constants in each layer a as ${}^ag_1$, ${}^ag_2$, …, ${}^a g_{10}$. In the equation above: the subscript 'D' labels Dark matter interactions, 'W' labels Weak fields, 'E' labels Electromagnetic fields, U labels U(4) Generation group fields, 'V' labels U(4) Layer group fields, U_D labels Dark U(4) Generation group fields, and V_D labels Dark U(4) Layer group fields. A_S labels the U(4) Species Group fields, Y labels the NORMAL Fermion field, and Y_D labels the DARK Fermion field. There are also spinor connection fields $B^{1\mu}$ and $B^{2\mu}$.

We define the sum over a and the components of the vector ${}^a\mathbf{A}_I{}^\mu(x)$ labeled with i = 1, 2 for each of the paired PseudoQuantum fields, by

$$ \mathbf{C}_I{}^\mu(x) = \Sigma_{a,i}\ {}^a\mathbf{A}_{Ii}{}^\mu(x) + \Sigma_i\ (g_{11}\mathbf{A}_S{}^{i\mu}(x) \qquad (5.2) $$

[78] This section has equations obtained from Blaha (2017d) and (2018e).

We begin by considering the case of one layer of vector bosons below omitting the ^{a} superscript. The generalization to four layers is straightforward.

Using the above definitions the *PseudoQuantum* covariant derivative of a 4-vector Z_μ is

$$D_\nu Z_\mu = (\partial_\nu + iF_\nu)Z_\mu - H^\sigma{}_{\nu\mu}Z_\sigma = [g^\sigma{}_\mu\partial_\nu + ig^\sigma{}_\mu F_\nu - H^\sigma{}_{\nu\mu}]Z_\sigma = [g^\sigma{}_\mu\partial_\nu + iD^\sigma{}_{\mu\nu}]Z_\sigma \quad (5.3)$$

where[79]

$$F^\mu = C_I{}^{1\mu}(x) + \mathbf{C}_I{}^{2\mu}(x) + B^{1\mu} + B^{2\mu} \quad (5.4)$$

and

$$H^\sigma{}_{\nu\mu} = \Gamma_{GR}{}^\sigma{}_{\nu\mu} + \Gamma_{GR}{}^{2\sigma}{}_{\nu\mu} \quad (5.5)$$
$$D^\sigma{}_{\mu\nu} = g^\sigma{}_\mu F_\nu + iH^\sigma{}_{\nu\mu}$$

where we have abstracted the complex part of the complex affine connection into the U(4) gauge field $A_S{}^\mu$. $H^\sigma{}_{\nu\mu}$ is the real-valued part of the complex affine connection.

We define the full vector gauge field covariant derivative to be

$$D^\mu = \partial^\mu - i(C_I{}^{1\mu}(x) + \mathbf{C}_I{}^{2\mu}(x)) \times \quad (5.5a)$$

Commutation relations of the vector fields in F_μ are implicit when the covariant derivative is applied to vectors and tensors such as Z_σ. This is indicated by the $\times$ symbol above.

5.2 The Curvature Tensor

The curvature tensor applied to a 4-vector Z_β is[80]

$$R'^\beta{}_{\sigma\nu\mu}Z_\beta = g^\alpha{}_\mu(\partial_\nu + iF_\nu)g^\beta{}_\sigma(\partial_\alpha + iF_\alpha)Z_\beta - H^\alpha{}_{\mu\nu}g^\beta{}_\sigma(\partial_\alpha + iF_\alpha)Z_\beta + H^\alpha{}_{\mu\nu}H^\beta{}_{\sigma\alpha}Z_\beta - g^\alpha{}_\mu(\partial_\nu + iF_\nu)H^\beta{}_{\sigma\alpha}Z_\beta - H^\gamma{}_{\nu\sigma}\{g^\alpha{}_\gamma(\partial_\mu + iF_\mu)Z_\alpha - H^\alpha{}_{\gamma\mu}Z_\alpha\} - \{\mu\leftrightarrow\nu\} \quad (5.6)$$

$$= ig^\beta{}_\sigma(\partial_\nu F_\mu - \partial_\mu F_\nu - i[F_\nu, F_\mu])Z_\beta + (\partial_\mu H^\beta{}_{\sigma\nu} - \partial_\nu H^\beta{}_{\sigma\mu} + H^\gamma{}_{\nu\sigma}H^\beta{}_{\gamma\mu} - H^\gamma{}_{\mu\sigma}H^\beta{}_{\gamma\nu})Z_\beta$$

$$= ig^\beta{}_\sigma(F_{E}{}^1{}_{\nu\mu} + F_{E}{}^2{}_{\nu\mu} + F_{W}{}^1{}_{\nu\mu} + F_{W}{}^2{}_{\nu\mu} + F_{DE}{}^1{}_{\nu\mu} + F_{DE}{}^2{}_{\nu\mu} + F_{DW}{}^1{}_{\nu\mu} + F_{DW}{}^2{}_{\nu\mu} + F_{SU(3)}{}^1{}_{\nu\mu} + F_{SU(3)}{}^2{}_{\nu\mu} + F_{DSU(3)}{}^1{}_{\nu\mu} + F_{DSU(3)}{}^2{}_{\nu\mu} + F_{U}{}^1{}_{\nu\mu} + F_{U}{}^2{}_{\nu\mu} + F_{V}{}^1{}_{\nu\mu} + F_{V}{}^2{}_{\nu\mu} + F_{UD}{}^1{}_{\nu\mu} + F_{UD}{}^2{}_{\nu\mu} + F_{VD}{}^1{}_{\nu\mu} + F_{VD}{}^2{}_{\nu\mu} + F_{Y}{}^1{}_{\nu\mu} + F_{Y}{}^2{}_{\nu\mu} + F_{YD}{}^1{}_{\nu\mu} + F_{YD}{}^2{}_{\nu\mu} + F_{S}{}^1{}_{\nu\mu} + F_{S}{}^2{}_{\nu\mu} + F_{B}{}^1{}_{\nu\mu} + F_{B}{}^2{}_{\nu\mu})Z_\beta + (\partial_\mu H^\beta{}_{\sigma\nu} - \partial_\nu H^\beta{}_{\sigma\mu} + H^\gamma{}_{\nu\sigma}H^\beta{}_{\gamma\mu} - H^\gamma{}_{\mu\sigma}H^\beta{}_{\gamma\nu})Z_\beta$$

---

[79] We will omit the insertion of the spinor coupling constants of the spinor connection $B^{1\mu}$ and $B^{2\mu}$ in the interests of simplifying expressions.

[80] **With an implicit summation over layers understood.**

$$= R'_E{}^{\beta}{}_{\sigma\nu\mu}Z_\beta + R'_{SU(2)}{}^{\beta}{}_{\sigma\nu\mu}Z_\beta + R'_{DE}{}^{\beta}{}_{\sigma\nu\mu}Z_\beta + R'_{DSU(2)}{}^{\beta}{}_{\sigma\nu\mu}Z_\beta + R'_{SU(3)}{}^{\beta}{}_{\sigma\nu\mu}Z_\beta + $$
$$+ R'_{DSU(3)}{}^{\beta}{}_{\sigma\nu\mu}Z_\beta + R'_U{}^{\beta}{}_{\sigma\nu\mu}Z_\beta + R'_V{}^{\beta}{}_{\sigma\nu\mu}Z_\beta + R'_{UD}{}^{\beta}{}_{\sigma\nu\mu}Z_\beta + R'_{VD}{}^{\beta}{}_{\sigma\nu\mu}Z_\beta +$$
$$+ R'_Y{}^{\beta}{}_{\sigma\nu\mu}Z_\beta + R'_{YD}{}^{\beta}{}_{\sigma\nu\mu}Z_\beta + R'_s{}^{\beta}{}_{\sigma\nu}Z_\beta + R'_B{}^{\beta}{}_{\sigma\nu}Z_\beta + R'_G{}^{\beta}{}_{\sigma\nu\mu}Z_\beta$$

where all $F_{\ldots}{}^1{}_{\nu\mu}$ and $F_{\ldots}{}^2{}_{\nu\mu}$ terms have summations over the four layers (see below) except the terms $F_S{}^1{}_{\nu\mu} + F_S{}^2{}_{\nu\mu} + F_A{}^2{}_{\nu\mu} + F_B{}^1{}_{\nu\mu} + F_B{}^2{}_{\nu\mu}$, and where[81]

$$\begin{aligned}
R'_{SU(3)}{}^{\beta}{}_{\sigma\nu\mu} &= ig^{\beta}{}_{\sigma}(F_{SU(3)}{}^1{}_{\nu\mu} + F_{SU(3)}{}^2{}_{\nu\mu}) \\
R'_{SU(2)}{}^{\beta}{}_{\sigma\nu\mu} &= ig^{\beta}{}_{\sigma}(F_W{}^1{}_{\nu\mu} + F_W{}^2{}_{\nu\mu}) \\
R'_E{}^{\beta}{}_{\sigma\nu\mu} &= ig^{\beta}{}_{\sigma}(F_E{}^1{}_{\nu\mu} + F_E{}^2{}_{\nu\mu}) \\
R'_U{}^{\beta}{}_{\sigma\nu\mu} &= ig^{\beta}{}_{\sigma}(F_U{}^1{}_{\nu\mu} + F_U{}^2{}_{\nu\mu}) \\
R'_V{}^{\beta}{}_{\sigma\nu\mu} &= ig^{\beta}{}_{\sigma}(F_V{}^1{}_{\nu\mu} + F_V{}^2{}_{\nu\mu}) \\
R'_Y{}^{\beta}{}_{\sigma\nu\mu} &= ig^{\beta}{}_{\sigma}(F_Y{}^1{}_{\nu\mu} + F_Y{}^2{}_{\nu\mu}) \\
R'_{DSU(3)}{}^{\beta}{}_{\sigma\nu\mu} &= ig^{\beta}{}_{\sigma}(F_{DSU(3)}{}^1{}_{\nu\mu} + F_{DSU(3)}{}^2{}_{\nu\mu}) \\
R'_{DSU(2)}{}^{\beta}{}_{\sigma\nu\mu} &= ig^{\beta}{}_{\sigma}(F_{DW}{}^1{}_{\nu\mu} + F_{DW}{}^2{}_{\nu\mu}) \\
R'_{DE}{}^{\beta}{}_{\sigma\nu\mu} &= ig^{\beta}{}_{\sigma}(F_{DE}{}^1{}_{\nu\mu} + F_{DE}{}^2{}_{\nu\mu}) \\
R'_{UD}{}^{\beta}{}_{\sigma\nu\mu} &= ig^{\beta}{}_{\sigma}(F_{UD}{}^1{}_{\nu\mu} + F_{UD}{}^2{}_{\nu\mu}) \\
R'_{VD}{}^{\beta}{}_{\sigma\nu\mu} &= ig^{\beta}{}_{\sigma}(F_{VD}{}^1{}_{\nu\mu} + F_{VD}{}^2{}_{\nu\mu}) \\
R'_{YD}{}^{\beta}{}_{\sigma\nu\mu} &= ig^{\beta}{}_{\sigma}(F_{YD}{}^1{}_{\nu\mu} + F_{YD}{}^2{}_{\nu\mu}) \\
R'_s{}^{\beta}{}_{\sigma\nu\mu} &= ig^{\beta}{}_{\sigma}(F_S{}^1{}_{\nu\mu} + F_S{}^2{}_{\nu\mu}) \\
R'_B{}^{\beta}{}_{\sigma\nu\mu} &= ig^{\beta}{}_{\sigma}(F_B{}^1{}_{\nu\mu} + F_B{}^2{}_{\nu\mu})
\end{aligned} \tag{5.7}$$

and

$$\begin{aligned}
R'_G{}^{\beta}{}_{\sigma\nu\mu} &= \partial_\mu H^{1\beta}{}_{\sigma\nu} - \partial_\nu H^{1\beta}{}_{\sigma\mu} + H^{1\gamma}{}_{\nu\sigma}H^{1\beta}{}_{\gamma\mu} - H^{1\gamma}{}_{\mu\sigma}H^{1\beta}{}_{\gamma\nu} + \partial_\mu H^{2\beta}{}_{\sigma\nu} - \partial_\nu H^{2\beta}{}_{\sigma\mu} + \\
&\quad + H^{2\gamma}{}_{\nu\sigma}H^{2\beta}{}_{\gamma\mu} - H^{2\gamma}{}_{\mu\sigma}H^{2\beta}{}_{\gamma\nu} + H^{1\gamma}{}_{\nu\sigma}H^{2\beta}{}_{\gamma\mu} - H^{1\gamma}{}_{\mu\sigma}H^{2\beta}{}_{\gamma\nu} + H^{2\gamma}{}_{\nu\sigma}H^{1\beta}{}_{\gamma\mu} - \Gamma^{2\gamma}{}_{\mu\sigma}\Gamma^{\beta}{}_{\gamma\nu} \\
&= R^{1\beta}{}_{\sigma\nu\mu} + R^{2\beta}{}_{\sigma\nu\mu}
\end{aligned} \tag{5.8}$$

with

$$\begin{aligned}
H^{\beta}{}_{\sigma\nu\mu} &= \partial_\mu H^{\beta}{}_{\sigma\nu} - \partial_\nu H^{\beta}{}_{\sigma\mu} + H^{\gamma}{}_{\nu\sigma}H^{\beta}{}_{\gamma\mu} - H^{\gamma}{}_{\mu\sigma}H^{\beta}{}_{\gamma\nu} \\
R^{1\beta}{}_{\sigma\nu\mu} &= \partial_\mu H^{1\beta}{}_{\sigma\nu} - \partial_\nu H^{1\beta}{}_{\sigma\mu} + H^{1\gamma}{}_{\nu\sigma}H^{1\beta}{}_{\gamma\mu} - H^{1\gamma}{}_{\mu\sigma}H^{1\beta}{}_{\gamma\nu} \\
R^{2\beta}{}_{\sigma\nu\mu p} &= \partial_\mu H^{2\beta}{}_{\sigma\nu} - \partial_\nu H^{2\beta}{}_{\sigma\mu} + H^{2\gamma}{}_{\nu\sigma}H^{2\beta}{}_{\gamma\mu} - H^{2\gamma}{}_{\mu\sigma}H^{2\beta}{}_{\gamma\nu} + \\
&\quad + H^{1\gamma}{}_{\nu\sigma}H^{2\beta}{}_{\gamma\mu} - H^{1\gamma}{}_{\mu\sigma}H^{2\beta}{}_{\gamma\nu} + H^{2\gamma}{}_{\nu\sigma}H^{1\beta}{}_{\gamma\mu} - H^{2\gamma}{}_{\mu\sigma}H^{1\beta}{}_{\gamma\nu}
\end{aligned} \tag{5.9}$$

and

$$\begin{aligned}
H^{1\sigma}{}_{\nu\mu} &= \Gamma_{GR}{}^{\sigma}{}_{\nu\mu} \\
H^{2\sigma}{}_{\nu\mu} &= \Gamma_{GR}{}^{2\sigma}{}_{\nu\mu}
\end{aligned} \tag{5.10}$$

and with summations over four layers indicated by Σ (Layer numbers on fields are not shown to avoid clutter.) As a result we have

[81] The B field is the General Relativistic spinor connection. Its effects are miniscule in physical situations except for extreme cases that have not as yet been encountered experimentally.

$$F_{SU(3)}{}^{1}{}_{\kappa\mu} = \Sigma\ \{\partial A_{SU(3)}{}^{1}{}_{\mu}/\partial x^{\kappa} - \partial A_{SU(3)}{}^{1}{}_{\kappa}/\partial x^{\mu} + ig_1[A_{SU(3)}{}^{1}{}_{\kappa}, A_{SU(3)}{}^{1}{}_{\mu}]\ \} \quad (5.11)$$
$$F_{W}{}^{1}{}_{\kappa\mu} = \Sigma\ \{\partial W^{1}{}_{\mu}/\partial x^{\kappa} - \partial W^{1}{}_{\kappa}/\partial x^{\mu} + ig_2[W^{1}{}_{\kappa}, W^{1}{}_{\mu}]\ \}$$
$$F_{E}{}^{1}{}_{\kappa\mu} = \Sigma\ \{\partial A_{E}{}^{1}{}_{\mu}/\partial x^{\kappa} - \partial A_{E}{}^{1}{}_{\kappa}/\partial x^{\mu}\ \}$$
$$F_{Y}{}^{1}{}_{\kappa\mu} = \Sigma\ \{\partial Y^{1}{}_{\mu}/\partial x^{\kappa} - \partial Y^{1}{}_{\kappa}/\partial x^{\mu}\ \}$$
$$F_{DW}{}^{1}{}_{\kappa\mu} = \Sigma\ \{\partial W_{D}{}^{1}{}_{\mu}/\partial x^{\kappa} - \partial W_{D}{}^{1}{}_{\kappa}/\partial x^{\mu} + ig_4[W_{D}{}^{1}{}_{\kappa}, W_{D}{}^{1}{}_{\mu}]\ \}$$
$$F_{DE}{}^{1}{}_{\kappa\mu} = \Sigma\ \{\partial A_{DE}{}^{1}{}_{\mu}/\partial x^{\kappa} - \partial A_{DE}{}^{1}{}_{\kappa}/\partial x^{\mu}\}$$
$$F_{DSU(3)}{}^{1}{}_{\kappa\mu} = \Sigma\ \{\partial A_{DSU(3)}{}^{1}{}_{\mu}/\partial x^{\kappa} - \partial A_{DSU(3)}{}^{1}{}_{\kappa}/\partial x^{\mu} + ig_8[A_{DSU(3)}{}^{1}{}_{\kappa}, A_{DSU(3)}{}^{1}{}_{\mu}]\ \}$$
$$F_{U}{}^{1}{}_{\kappa\mu} = \Sigma\ \{\partial U^{1}{}_{\mu}/\partial x^{\kappa} - \partial U^{1}{}_{\kappa}/\partial x^{\mu} + ig_6[U^{1}{}_{\kappa}, U^{1}{}_{\mu}]\ \}$$
$$F_{V}{}^{1}{}_{\kappa\mu} = \Sigma\ \{\partial V^{1}{}_{\mu}/\partial x^{\kappa} - \partial V^{1}{}_{\kappa}/\partial x^{\mu} + ig_7[V^{1}{}_{\kappa}, V^{1}{}_{\mu}]\ \}$$
$$F_{UD}{}^{1}{}_{\kappa\mu} \equiv F_{DU}{}^{1}{}_{\kappa\mu} = \Sigma\ \{\partial U_{D}{}^{1}{}_{\mu}/\partial x^{\kappa} - \partial U_{D}{}^{1}{}_{\kappa}/\partial x^{\mu} + ig_9[U_{D}{}^{1}{}_{\kappa}, U_{D}{}^{1}{}_{\mu}]\ \}$$
$$F_{VD}{}^{1}{}_{\kappa\mu} \equiv F_{DV}{}^{1}{}_{\kappa\mu} = \Sigma\ \{\partial V_{D}{}^{1}{}_{\mu}/\partial x^{\kappa} - \partial V_{D}{}^{1}{}_{\kappa}/\partial x^{\mu} + ig_{10}[V_{D}{}^{1}{}_{\kappa}, V_{D}{}^{1}{}_{\mu}]\ \}$$
$$F_{S}{}^{1}{}_{\kappa\mu} = \partial A_{S}{}^{1}{}_{\mu}/\partial x^{\kappa} - \partial A_{S}{}^{1}{}_{\kappa}/\partial x^{\mu} + ig_{11}[A_{S}{}^{1}{}_{\kappa}, A_{S}{}^{1}{}_{\mu}]$$
$$F_{YD}{}^{1}{}_{\kappa\mu} = \Sigma\ \{\partial Y_{D}{}^{1}{}_{\mu}/\partial x^{\kappa} - \partial Y_{D}{}^{1}{}_{\kappa}/\partial x^{\mu}\ \}$$
$$F_{B}{}^{1}{}_{\kappa\mu} = \partial B^{1}{}_{\mu}/\partial x^{\kappa} - \partial B^{1}{}_{\kappa}/\partial x^{\mu} + i[B^{1}{}_{\kappa}, B^{1}{}_{\mu}]$$

$$F_{SU(3)}{}^{2}{}_{\kappa\mu} = \Sigma\ \{\partial A_{SU(3)}{}^{2}{}_{\mu}/\partial x^{\kappa} - \partial A_{SU(3)}{}^{2}{}_{\kappa}/\partial x^{\mu} + ig_1[A_{SU(3)}{}^{2}{}_{\kappa}, A_{SU(3)}{}^{2}{}_{\mu}] + \quad (5.12)$$
$$+ ig_1[A_{SU(3)}{}^{1}{}_{\kappa}, A_{SU(3)}{}^{2}{}_{\mu}] + ig_1[A_{SU(3)}{}^{2}{}_{\kappa}, A_{SU(3)}{}^{1}{}_{\mu}]\}$$

$$F_{W}{}^{2}{}_{\kappa\mu} = \Sigma\ \{\partial W^{2}{}_{\mu}/\partial x^{\kappa} - \partial W^{2}{}_{\kappa}/\partial x^{\mu} + ig_2[W^{2}{}_{\kappa}, W^{2}{}_{\mu}] + ig_2[W^{1}{}_{\kappa}, W^{2}{}_{\mu}] + ig_2[W^{2}{}_{\kappa}, W^{1}{}_{\mu}]\ \}$$
$$F_{E}{}^{2}{}_{\kappa\mu} = \Sigma\ \{\partial A_{E}{}^{2}{}_{\mu}/\partial x^{\kappa} - \partial A_{E}{}^{2}{}_{\kappa}/\partial x^{\mu}\}$$
$$F_{Y}{}^{2}{}_{\kappa\mu} = \Sigma\ \{\partial Y^{2}{}_{\mu}/\partial x^{\kappa} - \partial Y^{2}{}_{\kappa}/\partial x^{\mu}\}$$

$$F_{DSU(3)}{}^{2}{}_{\kappa\mu} = \Sigma\ \{\partial A_{DSU(3)}{}^{2}{}_{\mu}/\partial x^{\kappa} - \partial A_{DSU(3)}{}^{2}{}_{\kappa}/\partial x^{\mu} + ig_8[A_{DSU(3)}{}^{2}{}_{\kappa}, A_{DSU(3)}{}^{2}{}_{\mu}] +$$
$$+ ig_8[A_{DSU(3)}{}^{1}{}_{\kappa}, A_{DSU(3)}{}^{2}{}_{\mu}] + ig_8[A_{DSU(3)}{}^{2}{}_{\kappa}, A_{DSU(3)}{}^{1}{}_{\mu}]\}$$

$$F_{DW}{}^{2}{}_{\kappa\mu} = \Sigma\ \{\partial W_{D}{}^{2}{}_{\mu}/\partial x^{\kappa} - \partial W_{D}{}^{2}{}_{\kappa}/\partial x^{\mu} + ig_4[W_{D}{}^{2}{}_{\kappa}, W_{D}{}^{2}{}_{\mu}] + ig_4[W_{D}{}^{1}{}_{\kappa}, W_{D}{}^{2}{}_{\mu}] +$$
$$+ ig_4[W_{D}{}^{2}{}_{\kappa}, W_{D}{}^{1}{}_{\mu}]\}$$

$$F_{DE}{}^{2}{}_{\kappa\mu} = \Sigma\ \{\partial A_{DE}{}^{2}{}_{\mu}/\partial x^{\kappa} - \partial A_{DE}{}^{2}{}_{\kappa}/\partial x^{\mu}\}$$

$$F_{U}{}^{2}{}_{\kappa\mu} = \Sigma\ \{\partial U^{2}{}_{\mu}/\partial x^{\kappa} - \partial U^{2}{}_{\kappa}/\partial x^{\mu} + ig_6[U^{2}{}_{\kappa}, U^{2}{}_{\mu}] + ig_6[U^{1}{}_{\kappa}, U^{2}{}_{\mu}] + ig_6[U^{2}{}_{\kappa}, U^{1}{}_{\mu}]\}$$

$$F_{V}{}^{2}{}_{\kappa\mu} = \Sigma\ \{\partial V^{2}{}_{\mu}/\partial x^{\kappa} - \partial V^{2}{}_{\kappa}/\partial x^{\mu} + ig_7[V^{2}{}_{\kappa}, V^{2}{}_{\mu}] + ig_7[V^{1}{}_{\kappa}, V^{2}{}_{\mu}] + ig_7[V^{2}{}_{\kappa}, V^{1}{}_{\mu}]\}$$

$$F_{UD}{}^{2}{}_{\kappa\mu} = \Sigma\ \{\partial U_{D}{}^{2}{}_{\mu}/\partial x^{\kappa} - \partial U_{D}{}^{2}{}_{\kappa}/\partial x^{\mu} + ig_9[U_{D}{}^{2}{}_{\kappa}, U_{D}{}^{2}{}_{\mu}] + ig_9[U_{D}{}^{1}{}_{\kappa}, U_{D}{}^{2}{}_{\mu}] +$$

$$+ ig_6[U_D{}^2{}_\varkappa, U_D{}^1{}_\mu]\}$$
$$\equiv F_{DU}{}^2{}_{\varkappa\mu}$$
$$F_{VD}{}^2{}_{\varkappa\mu} = \Sigma\ \{\partial V_D{}^2{}_\mu/\partial x^\varkappa - \partial V_D{}^2{}_\varkappa/\partial x^\mu + ig_7[V_D{}^2{}_\varkappa, V_D{}^2{}_\mu] + ig_7[V_D{}^1{}_\varkappa, V_D{}^2{}_\mu] +$$
$$+ ig_7[V_D{}^2{}_\varkappa, V_D{}^1{}_\mu]\}$$
$$\equiv F_{DV}{}^2{}_{\varkappa\mu}$$
$$F_{YD}{}^2{}_{\varkappa\mu} = \Sigma\ \{\partial Y_D{}^2{}_\mu/\partial x^\varkappa - \partial Y_D{}^2{}_\varkappa/\partial x^\mu\}$$

$$F_S{}^2{}_{\kappa\mu} = \partial A_S{}^2{}_\mu/\partial x^\kappa - \partial A_S{}^2{}_\kappa/\partial x^\mu + ig_8[A_S{}^2{}_\kappa, A_S{}^2{}_\mu] + ig_{11}[A_S{}^1{}_\kappa, A_S{}^2{}_\mu] + ig_{11}[A_S{}^2{}_\kappa, A_S{}^1{}_\mu]$$

$$F_B{}^2{}_{\varkappa\mu} = \partial B^2{}_\mu/\partial x^\varkappa - \partial B^2{}_\varkappa/\partial x^\mu + i[B^2{}_\mu, B^2{}_\varkappa] + i[B^1{}_\mu, B^2{}_\varkappa] + i[B^2{}_\mu, B^1{}_\varkappa]$$

Note that $R'^\beta{}_{\sigma\nu\mu}$ factorizes into a

$$[U(1)\otimes SU(2)\otimes U(1)\otimes SU(2)\otimes SU(3)\otimes SU(3)\otimes U(4)\otimes U(4)\otimes U(1)\otimes U(1)]^4$$

part and a Riemann-Christoffel Gravitational curvature tensor part. For later use in defining a lagrangian we define

$$\begin{aligned} R'^\beta{}_{\sigma\nu\mu} = {} & R'_E{}^{1\beta}{}_{\sigma\nu\mu} + R'_E{}^{2\beta}{}_{\sigma\nu\mu} + R'_{SU(2)}{}^{1\beta}{}_{\sigma\nu\mu} + R'_{SU(2)}{}^{2\beta}{}_{\sigma\nu\mu} + R'_{DE}{}^{1\beta}{}_{\sigma\nu\mu} + R'_{DE}{}^{2\beta}{}_{\sigma\nu\mu} + \\ & + R'_{DSU(2)}{}^{1\beta}{}_{\sigma\nu\mu} + R'_{DSU(2)}{}^{2\beta}{}_{\sigma\nu\mu} + R'_{SU(3)}{}^{1\beta}{}_{\sigma\nu\mu} + R'_{SU(3)}{}^{2\beta}{}_{\sigma\nu\mu} + R'_{DSU(3)}{}^{1\beta}{}_{\sigma\nu\mu} + \\ & + R'_{DSU(3)}{}^{2\beta}{}_{\sigma\nu\mu} + R'_U{}^{1\beta}{}_{\sigma\nu\mu} + R'_U{}^{2\beta}{}_{\sigma\nu\mu} + R'_V{}^{1\beta}{}_{\sigma\nu\mu} + R'_V{}^{2\beta}{}_{\sigma\nu\mu} + R'_{DU}{}^{1\beta}{}_{\sigma\nu\mu} + \\ & + R'_{DU}{}^{2\beta}{}_{\sigma\nu\mu} + R'_{DV}{}^{1\beta}{}_{\sigma\nu\mu} + R'_{DV}{}^{2\beta}{}_{\sigma\nu\mu} + R'_Y{}^{2\beta}{}_{\sigma\nu\mu} + R'_Y{}^{1\beta}{}_{\sigma\nu\mu} + R'_{DY}{}^{1\beta}{}_{\sigma\nu\mu} + \\ & + R'_{DY}{}^{2\beta}{}_{\sigma\nu\mu} + R'_S{}^{1\beta}{}_{\sigma\nu\mu} + R'_S{}^{2\beta}{}_{\sigma\nu\mu} + R'_B{}^{1\beta}{}_{\sigma\nu\mu} + R'_B{}^{2\beta}{}_{\sigma\nu\mu} + R^{1\beta}{}_{\sigma\nu\mu} + R^{2\beta}{}_{\sigma\nu\mu} \end{aligned} \quad (5.13)$$

where

$$\begin{aligned} R'_E{}^{1\beta}{}_{\sigma\nu\mu} &= ig^\beta{}_\sigma F_E{}^1{}_{\nu\mu} \\ R'_E{}^{2\beta}{}_{\sigma\nu\mu} &= ig^\beta{}_\sigma F_{DE}{}^2{}_{\nu\mu} \\ R'_{DE}{}^{1\beta}{}_{\sigma\nu\mu} &= ig^\beta{}_\sigma F_E{}^1{}_{\nu\mu} \\ R'_{DE}{}^{2\beta}{}_{\sigma\nu\mu} &= ig^\beta{}_\sigma F_{DE}{}^2{}_{\nu\mu} \\ R'_{SU(2)}{}^{1\beta}{}_{\sigma\nu\mu} &= ig^\beta{}_\sigma F_W{}^1{}_{\nu\mu} \\ R'_{SU(2)}{}^{2\beta}{}_{\sigma\nu\mu} &= ig^\beta{}_\sigma F_{DW}{}^2{}_{\nu\mu} \\ R'_{DSU(2)}{}^{1\beta}{}_{\sigma\nu\mu} &= ig^\beta{}_\sigma F_W{}^1{}_{\nu\mu} \\ R'_{DSU(2)}{}^{2\beta}{}_{\sigma\nu\mu} &= ig^\beta{}_\sigma F_{DW}{}^2{}_{\nu\mu} \\ R'_{SU(3)}{}^{1\beta}{}_{\sigma\nu\mu} &= ig^\beta{}_\sigma F_{SU(3)}{}^1{}_{\nu\mu} \\ R'_{SU(3)}{}^{2\beta}{}_{\sigma\nu\mu} &= ig^\beta{}_\sigma F_{SU(3)}{}^2{}_{\nu\mu} \\ R'_{DSU(3)}{}^{1\beta}{}_{\sigma\nu\mu} &= ig^\beta{}_\sigma F_{DSU(3)}{}^1{}_{\nu\mu} \\ R'_{DSU(3)}{}^{2\beta}{}_{\sigma\nu\mu} &= ig^\beta{}_\sigma F_{DSU(3)}{}^2{}_{\nu\mu} \\ R'_U{}^{1\beta}{}_{\sigma\nu\mu} &= ig^\beta{}_\sigma F_U{}^1{}_{\nu\mu} \\ R'_U{}^{2\beta}{}_{\sigma\nu\mu} &= ig^\beta{}_\sigma F_U{}^2{}_{\nu\mu} \\ R'_V{}^{1\beta}{}_{\sigma\nu\mu} &= ig^\beta{}_\sigma F_V{}^1{}_{\nu\mu} \\ R'_V{}^{2\beta}{}_{\sigma\nu\mu} &= ig^\beta{}_\sigma F_V{}^2{}_{\nu\mu} \end{aligned} \quad (5.14)$$

$$R'_{DU}{}^{1\beta}{}_{\sigma\nu\mu} = ig^{\beta}{}_{\sigma}F_{DU}{}^{1}{}_{\nu\mu}$$
$$R'_{DU}{}^{2\beta}{}_{\sigma\nu\mu} = ig^{\beta}{}_{\sigma}F_{DU}{}^{2}{}_{\nu\mu}$$
$$R'_{DV}{}^{1\beta}{}_{\sigma\nu\mu} = ig^{\beta}{}_{\sigma}F_{DV}{}^{1}{}_{\nu\mu}$$
$$R'_{DV}{}^{2\beta}{}_{\sigma\nu\mu} = ig^{\beta}{}_{\sigma}F_{DV}{}^{2}{}_{\nu\mu}$$
$$R'_{Y}{}^{1\beta}{}_{\sigma\nu\mu} = ig^{\beta}{}_{\sigma}F_{Y}{}^{1}{}_{\nu\mu}$$
$$R'_{Y}{}^{2\beta}{}_{\sigma\nu\mu} = ig^{\beta}{}_{\sigma}F_{Y}{}^{2}{}_{\nu\mu}$$
$$R'_{DY}{}^{1\beta}{}_{\sigma\nu\mu} = ig^{\beta}{}_{\sigma}F_{DY}{}^{1}{}_{\nu\mu}$$
$$R'_{DY}{}^{2\beta}{}_{\sigma\nu\mu} = ig^{\beta}{}_{\sigma}F_{DY}{}^{2}{}_{\nu\mu}$$
$$R'_{S}{}^{1\beta}{}_{\sigma\nu\mu} = ig^{\beta}{}_{\sigma}F_{S}{}^{1}{}_{\nu\mu}$$
$$R'_{S}{}^{2\beta}{}_{\sigma\nu\mu} = ig^{\beta}{}_{\sigma}F_{S}{}^{2}{}_{\nu\mu}$$
$$R'_{B}{}^{1\beta}{}_{\sigma\nu\mu} = ig^{\beta}{}_{\sigma}B^{1}{}_{\nu\mu}$$
$$R'_{B}{}^{2\beta}{}_{\sigma\nu\mu} = ig^{\beta}{}_{\sigma}B^{2}{}_{\nu\mu}$$

The total Ricci tensor is

$$R'_{\sigma\mu} = R'^{\beta}{}_{\sigma\beta\mu} \tag{5.15}$$

$$\begin{aligned}
&= iF_{E}{}^{1}{}_{\sigma\mu} + iF_{E}{}^{2}{}_{\sigma\mu} + iF_{W}{}^{1}{}_{\sigma\mu} + iF_{W}{}^{2}{}_{\sigma\mu} + iF_{DE}{}^{1}{}_{\sigma\mu} + iF_{DE}{}^{2}{}_{\sigma\mu} + iF_{DW}{}^{1}{}_{\sigma\mu} + iF_{DW}{}^{2}{}_{\sigma\mu} + iF_{SU(3)}{}^{1}{}_{\sigma\mu} + iF_{SU(3)}{}^{2}{}_{\sigma\mu} + \\
&\quad + iF_{DSU(3)}{}^{1}{}_{\sigma\mu} + iF_{DSU(3)}{}^{2}{}_{\sigma\mu} + iF_{U}{}^{1}{}_{\sigma\mu} + iF_{U}{}^{2}{}_{\sigma\mu} + iF_{V}{}^{1}{}_{\sigma\mu} + iF_{V}{}^{2}{}_{\sigma\mu} + iF_{DU}{}^{1}{}_{\sigma\mu} + iF_{DU}{}^{2}{}_{\sigma\mu} + iF_{DV}{}^{1}{}_{\sigma\mu} + \\
&\quad + iF_{DV}{}^{2}{}_{\sigma\mu} + iF_{Y}{}^{1}{}_{\sigma\mu} + iF_{Y}{}^{2}{}_{\sigma\mu} + iF_{DY}{}^{1}{}_{\sigma\mu} + iF_{DY}{}^{2}{}_{\sigma\mu} + iF_{S}{}^{1}{}_{\sigma\mu} + iF_{S}{}^{2}{}_{\sigma\mu} + iF_{B}{}^{1}{}_{\sigma\mu} + iF_{B}{}^{2}{}_{\sigma\mu} + \\
&\quad + \partial_{\mu}H^{1\beta}{}_{\sigma\beta} - \partial_{\beta}H^{1\beta}{}_{\sigma\mu} + H^{1\gamma}{}_{\beta\sigma}H^{1\beta}{}_{\gamma\mu} - H^{1\gamma}{}_{\mu\sigma}H^{1\beta}{}_{\gamma\beta} + \\
&\quad + \partial_{\mu}H^{2\beta}{}_{\sigma\beta} - \partial_{\beta}H^{2\beta}{}_{\sigma\mu} + H^{2\gamma}{}_{\beta\sigma}H^{2\beta}{}_{\gamma\mu} - H^{2\gamma}{}_{\mu\sigma}H^{2\beta}{}_{\gamma\beta} + H^{1\gamma}{}_{\beta\sigma}H^{2\beta}{}_{\gamma\mu} - H^{1\gamma}{}_{\mu\sigma}H^{2\beta}{}_{\gamma\beta} + \\
&\quad + H^{2\gamma}{}_{\beta\sigma}H^{1\beta}{}_{\gamma\mu} - H^{2\gamma}{}_{\mu\sigma}H^{1\beta}{}_{\gamma\beta}
\end{aligned}$$

$$\begin{aligned}
&= R'_{E}{}^{1}{}_{\sigma\mu} + R'_{E}{}^{2}{}_{\sigma\mu} + R'_{SU(2)}{}^{1}{}_{\sigma\mu} + R'_{SU(2)}{}^{2}{}_{\sigma\mu} + R'_{DE}{}^{1}{}_{\sigma\mu} + R'_{DE}{}^{2}{}_{\sigma\mu} + R'_{DSU(2)}{}^{1}{}_{\sigma\mu} + R'_{DSU(2)}{}^{2}{}_{\sigma\mu} + \\
&\quad + R'_{SU(3)}{}^{1}{}_{\sigma\mu} + R'_{SU(3)}{}^{2}{}_{\sigma\mu} + R'_{U}{}^{1}{}_{\sigma\mu} + R'_{U}{}^{2}{}_{\sigma\mu} + R'_{V}{}^{1}{}_{\sigma\mu} + R'_{V}{}^{2}{}_{\sigma\mu} + R'_{Y}{}^{1}{}_{\sigma\mu} + R'_{Y}{}^{2}{}_{\sigma\mu} + \\
&\quad + R'_{DSU(3)}{}^{1}{}_{\sigma\mu} + R'_{DSU(3)}{}^{2}{}_{\sigma\mu} + \\
&\quad + R'_{DU}{}^{1}{}_{\sigma\mu} + R'_{DU}{}^{2}{}_{\sigma\mu} + R'_{DV}{}^{1}{}_{\sigma\mu} + R'_{DV}{}^{2}{}_{\sigma\mu} + R'_{DY}{}^{1}{}_{\sigma\mu} + R'_{DY}{}^{2}{}_{\sigma\mu} + R'_{S}{}^{1}{}_{\sigma\mu} + R'_{S}{}^{2}{}_{\sigma\mu} + \\
&\quad + R'_{A}{}^{1\beta}{}_{\sigma\beta\mu} + R'_{A}{}^{2\beta}{}_{\sigma\beta\mu} + R'_{B}{}^{1\beta}{}_{\sigma\beta\mu} + R'_{B}{}^{2\beta}{}_{\sigma\beta\mu} + R^{1}{}_{\sigma\mu} + R^{2}{}_{\sigma\mu} \\
&= R'^{1}{}_{\sigma\mu} + R'^{2}{}_{\sigma\mu}
\end{aligned}$$

where

$$\begin{aligned}
R'^{1}{}_{\sigma\mu} &= R'_{E}{}^{1}{}_{\sigma\mu} + R'_{SU(2)}{}^{1}{}_{\sigma\mu} + R'_{DE}{}^{1}{}_{\sigma\mu} + R'_{DSU(2)}{}^{1}{}_{\sigma\mu} + R'_{SU(3)}{}^{1}{}_{\sigma\mu} + R'_{U}{}^{1}{}_{\sigma\mu} + R'_{V}{}^{1}{}_{\sigma\mu} + \\
&\quad + R'_{Y}{}^{1}{}_{\sigma\mu} + R'_{DY}{}^{1}{}_{\sigma\mu} + R'_{DSU(3)}{}^{1}{}_{\sigma\mu} + R'_{DU}{}^{1}{}_{\sigma\mu} + R'_{DV}{}^{1}{}_{\sigma\mu} + R'_{S}{}^{1}{}_{\sigma\mu} + R'_{B}{}^{1\beta}{}_{\sigma\beta\mu} + R^{1}{}_{\sigma\mu}
\end{aligned} \tag{5.16}$$

$$\begin{aligned}
R'^{2}{}_{\sigma\mu} &= R'_{E}{}^{2}{}_{\sigma\mu} + R'_{SU(2)}{}^{2}{}_{\sigma\mu} + R'_{DE}{}^{2}{}_{\sigma\mu} + R'_{DSU(2)}{}^{2}{}_{\sigma\mu} + R'_{SU(3)}{}^{2}{}_{\sigma\mu} + R'_{U}{}^{2}{}_{\sigma\mu} + R'_{V}{}^{2}{}_{\sigma\mu} + R'_{Y}{}^{2}{}_{\sigma\mu} + \\
&\quad + R'_{DY}{}^{2}{}_{\sigma\mu} + R'_{DSU(3)}{}^{2}{}_{\sigma\mu} + R'_{DU}{}^{2}{}_{\sigma\mu} + R'_{DV}{}^{2}{}_{\sigma\mu} + R'_{S}{}^{2}{}_{\sigma\mu} + R'_{B}{}^{2\beta}{}_{\sigma\beta\mu} + R^{2}{}_{\sigma\mu}
\end{aligned} \tag{5.17}$$

with

$$R'_{E}{}^{1}{}_{\sigma\mu} = iF_{E}{}^{1}{}_{\sigma\mu} \tag{5.18}$$
$$R'_{E}{}^{2}{}_{\sigma\mu} = iF_{E}{}^{2}{}_{\sigma\mu}$$

$$R'_{SU(2)}{}^{1}{}_{\sigma\mu} = iF_{W}{}^{1}{}_{\sigma\mu}$$
$$R'_{SU(2)}{}^{2}{}_{\sigma\mu} = iF_{W}{}^{2}{}_{\sigma\mu}$$
$$R'_{DE}{}^{1}{}_{\sigma\mu} = iF_{DE}{}^{1}{}_{\sigma\mu}$$
$$R'_{DE}{}^{2}{}_{\sigma\mu} = iF_{DE}{}^{2}{}_{\sigma\mu}$$
$$R'_{DSU(2)}{}^{1}{}_{\sigma\mu} = iF_{DW}{}^{1}{}_{\sigma\mu}$$
$$R'_{DSU(2)}{}^{2}{}_{\sigma\mu} = iF_{DW}{}^{2}{}_{\sigma\mu}$$
$$R'_{SU(3)}{}^{1}{}_{\sigma\mu} = iF_{SU(3)}{}^{1}{}_{\sigma\mu}$$
$$R'_{SU(3)}{}^{2}{}_{\sigma\mu} = iF_{SU(3)}{}^{2}{}_{\sigma\mu}$$
$$R'_{U}{}^{1}{}_{\sigma\mu} = iF_{U}{}^{1}{}_{\sigma\mu}$$
$$R'_{U}{}^{2}{}_{\sigma\mu} = iF_{U}{}^{2}{}_{\sigma\mu}$$
$$R'_{V}{}^{1}{}_{\sigma\mu} = iF_{V}{}^{1}{}_{\sigma\mu}$$
$$R'_{V}{}^{2}{}_{\sigma\mu} = iF_{V}{}^{2}{}_{\sigma\mu}$$
$$R'_{Y}{}^{1}{}_{\sigma\mu} = iF_{Y}{}^{1}{}_{\sigma\mu}$$
$$R'_{Y}{}^{2}{}_{\sigma\mu} = iF_{Y}{}^{2}{}_{\sigma\mu}$$
$$R'_{DSU(3)}{}^{1}{}_{\sigma\mu} = iF_{DSU(3)}{}^{1}{}_{\sigma\mu}$$
$$R'_{DSU(3)}{}^{2}{}_{\sigma\mu} = iF_{DSU(3)}{}^{2}{}_{\sigma\mu}$$
$$R'_{DU}{}^{1}{}_{\sigma\mu} = iF_{DU}{}^{1}{}_{\sigma\mu}$$
$$R'_{DU}{}^{2}{}_{\sigma\mu} = iF_{DU}{}^{2}{}_{\sigma\mu}$$
$$R'_{DV}{}^{1}{}_{\sigma\mu} = iF_{DV}{}^{1}{}_{\sigma\mu}$$
$$R'_{DV}{}^{2}{}_{\sigma\mu} = iF_{DV}{}^{2}{}_{\sigma\mu}$$
$$R'_{DY}{}^{1}{}_{\sigma\mu} = iF_{DY}{}^{1}{}_{\sigma\mu}$$
$$R'_{DY}{}^{2}{}_{\sigma\mu} = iF_{DY}{}^{2}{}_{\sigma\mu}$$
$$R'_{S}{}^{1}{}_{\sigma\mu} = iF_{S}{}^{1}{}_{\sigma\mu}$$
$$R'_{S}{}^{2}{}_{\sigma\mu} = iF_{S}{}^{2}{}_{\sigma\mu}$$
$$R'_{B}{}^{1}{}_{\sigma\mu} = iF_{B}{}^{1}{}_{\sigma\mu}$$
$$R'_{B}{}^{2}{}_{\sigma\mu} = iF_{B}{}^{2}{}_{\sigma\mu}$$

with the further definition of $R''^{1}{}_{\sigma\mu}$ and $R''^{2}{}_{\sigma\mu}$:

$$\begin{aligned} R''^{1}{}_{\sigma\mu} &= R'_{SU(3)}{}^{1}{}_{\sigma\mu} + R'_{DSU(3)}{}^{1}{}_{\sigma\mu} + R^{1}{}_{\sigma\mu} \\ R''^{2}{}_{\sigma\mu} &= R'_{SU(3)}{}^{2}{}_{\sigma\mu} + R'_{DSU(3)}{}^{2}{}_{\sigma\mu} + R^{2}{}_{\sigma\mu} \end{aligned} \tag{5.19}$$

$R'^{1}{}_{\sigma\mu}$ is the Ricci tensor. An additional Ricci-like tensor is

$$H_{\sigma_\mu} = H^{\beta}{}_{\sigma\beta\mu} \tag{5.20}$$

The curvature scalar is

$$\begin{aligned} R' = g^{\sigma\mu}R'_{\sigma\mu} = &+ \partial^{\sigma}H^{1\beta}{}_{\sigma\beta} - \partial_{\beta}H^{1\beta}{}_{\sigma}{}^{\sigma} + H^{1\gamma}{}_{\beta\sigma}H^{1\beta}{}_{\gamma}{}^{\sigma} - H^{1\gamma}{}_{\mu\sigma}H^{1\beta}{}_{\gamma\beta} + \partial^{\sigma}H^{2\beta}{}_{\sigma\beta} - \partial_{\beta}H^{2\beta}{}_{\sigma}{}^{\sigma} + \\ &+ H^{2\gamma}{}_{\beta\sigma}H^{2\beta}{}_{\gamma}{}^{\sigma} - H^{2\gamma\sigma}{}_{\sigma}H^{2\beta}{}_{\gamma\beta} + H^{1\gamma}{}_{\beta\sigma}H^{2\beta}{}_{\gamma}{}^{\sigma} - H^{1\gamma\sigma}{}_{\sigma}H^{2\beta}{}_{\gamma\beta} + H^{2\gamma}{}_{\beta\sigma}H^{1\beta}{}_{\gamma}{}^{\sigma} - \\ &- H^{2\gamma\sigma}{}_{\sigma}H^{1\beta}{}_{\gamma\beta} \end{aligned} \tag{5.21}$$

$$= g^{\sigma\mu}(R^{1\beta}{}_{\sigma\beta\mu} + R^{2\beta}{}_{\sigma\beta\mu})$$

5.3 Vector Boson and Graviton Lagrangian Terms

We choose the vector boson and gravitational part of the lagrangian[82] of the Unified SuperStandard Theory (with the Higgs sector and the Faddeev-Popov terms gauge sector not displayed here) to be:

$$\mathfrak{L} = \text{Tr}\sqrt{g}[MD_\nu R''^{1}{}_{\sigma\mu}D^\nu R''^{2\sigma\mu} + aR'^{1}{}_{\sigma\mu}R'^{2\sigma\mu} + bR' + cg^{\sigma\mu}g^{2}{}_{\sigma\mu} + c'g^{2\sigma\mu}g^{2}{}_{\sigma\mu} - \\ - dA_{SU(3)}{}^{2}{}_{\mu}A_{SU(3)}{}^{2\mu}] \qquad (5.22)$$

with a sum over layers understood, where D_ν is given by eq. 5.5a, where M, a, b, c, c', and d are constants, and $R''^{i}{}_{\sigma\mu}$ for i = 1, 2 determined above.[83]

This higher derivative lagrangian maintains the locality of the theory but does entail a modest modification in the derivation of the Euler-Lagrange equations of motion. It also requires the use of principal value propagators rather than ordinary Feynman propagators for gluon and graviton interactions. Thus the Strong Interaction sector, and the Gravitation sector are Action-at-a-Distance theories that are similar in spirit to Wheeler-Feynman Electrodynamics. The two U(1) Electromagnetic sectors, the Generation group U(4) gauge field sector, the Layer group U(4) gauge field sector, the two SU2) Weak sectors, the U(4) A_S gauge field sector, the spinor connection sector, and the Θ-interaction sector may, or may not, be Action-at-a-Distance fields. They are not constrained to be Action-at-a-Distance by the present considerations.

Since we wish to apply our theory cosmologically, and within hadrons, where the gravitational spinor connections are negligible due to the smallness of the

---

[82] The rationale for this choice is 1) to obtain the known Stanard Model interactions, 2) to obtain a canonical formulation for this higher derivative theory, and 3) to introduce higher derivative terms that yield quark confinement and a theory of gravity that accounts for known deviations from Newtonian gravity such as described by MoND. See Blaha (2019g) and (2018e) and earlier books for details on these points.

[83] One may ask why $R''^{1}{}_{\sigma\mu}$ and $R''^{2}{}_{\sigma\mu}$ appear in the first term of the lagrangian, and not other interaction terms. We believe the primary reason is: "The extended vierbein $1^{\mu ai}(x)$ can be viewed as located at a point in a higher dimensional complex-valued space.

$$1^{\mu ai}(x) = (\partial\xi_X{}^{ai}(x)/\partial x_\mu)_{X=h(x)}$$

where $\xi_X{}^{ai}$ is a set of locally inertial coordinates located at point X, and x = h(x) is a 4-dimensional point in a tangent subspace of the higher dimensional space:

$$X = h(x)$$

The relation between complex 4-dimensional coordinates x and the higher dimensional coordinates X is an embedding of a 4-dimensional surface within the higher dimensional complex space when account is taken of the range of possible x values. We have considered such embeddings in Blaha (2015a), and in earlier books, and developed a theory of a higher dimensional complex-valued space (the *Megaverse*) that contains our universe and probably many other universes." Thus SU(3) and Gravitation have a special role in our particle dynamics based on geometry. The second reason is the common feature of color SU(3) and real-valued General Relativity is that they are the only interactions that do not participate in 'rotations of interactions' as described earlier and in chapter 31 of Blaha (2017b). The third, practical reason is the experimental reality that the Strong Interaction and Gravitation are known to have 'anomalous' features that will be seen to be remedied by these insertions while the other interactions are 'conventional.'

gravitational constant G and the 'smallness' of Gravitational B spinor connection effects on the cosmological scale, we set $F^1{}_{\nu\mu} = F^2{}_{\nu\mu} = 0$ and find[84]

$$\mathcal{L} = \text{Tr}\ \sqrt{g}[MD_\nu(R'^1{}_{SU(3)\sigma\mu} + R'^1{}_{DSU(3)\sigma\mu} + R^1{}_{\sigma\mu})D^\nu(R'_{SU(3)}{}^{2\sigma\mu} + R'_{DSU(3)}{}^{2\sigma\mu} + R^{2\sigma\mu}) + + aR'^1{}_{\sigma\mu}R'^{2\sigma\mu} + bR' + cg^{\sigma\mu}g^2{}_{\sigma\mu} + c'g^{2\sigma\mu}g^2{}_{\sigma\mu} - dA_{SU(3)}{}^2{}_\mu A_{SU(3)}{}^{2\mu}] \tag{5.23}$$

Since there are no strong interaction fields in 'empty' space and gravity is negligible within hadrons,[85] we can drop the interaction terms between the Strong interaction and the Gravity interaction. However, we cannot drop the interaction terms amongst Electromagnetism, the Weak interaction, the Strong Interaction, the Generation groups U(4) interactions, the U(4) Layer groups interactions, and the U(4) Species group interaction – within, and between, hadrons. The interaction terms between Electromagnetism and Gravitation are important cosmologically.

The above lagrangian terms can therefore be expressed as:[86]

$$\mathcal{L} = \mathcal{L}_E + \mathcal{L}_{SU(2)} + \mathcal{L}_{DE} + \mathcal{L}_{DSU(2)} + \mathcal{L}_{SU(3)} + \mathcal{L}_{DSU(3)} + \mathcal{L}_U + \mathcal{L}_V + \mathcal{L}_{DU} + \mathcal{L}_{DV} + \mathcal{L}_Y + + \mathcal{L}_{DY} + \mathcal{L}_S + + \mathcal{L}_G + \mathcal{L}_{int} \tag{5.24}$$

where taking traces of $\mathcal{L}$s terms is understood, and with coupling constants not displayed below to avoid clutter,

$$\mathcal{L}_E = \text{Tr}\ \sqrt{g}\{M\{[\partial_\nu + i(A_E{}^1{}_\nu + A_E{}^2{}_\nu)]F^1{}_{E\sigma\mu}[\partial^\nu + i(A_E{}^{1\nu} + A_E{}^{2\nu})]F^2{}_E{}^{\sigma\mu}\} + aF_E{}^1{}_{\sigma\mu}F_E{}^{2\sigma\mu}\} \tag{5.25}$$

$$\mathcal{L}_{SU(2)} = \text{Tr}\ \sqrt{g}[aF_W{}^1{}_{\sigma\mu}F_W{}^{2\sigma\mu}]$$

$$\mathcal{L}_Y = \text{Tr}\ \sqrt{g}\{M\{[\partial_\nu + i(Y^1{}_\nu + Y^2{}_\nu)]F^1{}_{Y\sigma\mu}[\partial^\nu + i(Y^{1\nu} + Y^{2\nu})]F^2{}_Y{}^{\sigma\mu}\} + aF_Y{}^1{}_{\sigma\mu}F_Y{}^{2\sigma\mu}\}$$

$$\mathcal{L}_{DY} = \text{Tr}\ \sqrt{g}\{M\{[\partial_\nu + i(Y_D{}^1{}_\nu + Y_D{}^2{}_\nu)]F^1{}_{DY\sigma\mu}[\partial^\nu + i(Y_D{}^{1\nu} + Y_D{}^{2\nu})]F_{Dy}{}^{2\sigma\mu}\} + + aF_{DY}{}^1{}_{\sigma\mu}F_{DY}{}^{2\sigma\mu}\}$$

$$\mathcal{L}_{DE} = \text{Tr}\ \sqrt{g}\{M\{[\partial_\nu + i(A_{DE}{}^1{}_\nu + A_{DE}{}^2{}_\nu)]F^1{}_{DE\sigma\mu}[\partial^\nu + i(A_{DE}{}^{1\nu} + A_{DE}{}^{2\nu})]F_{DE}{}^{2\sigma\mu}\} + + aF_{DE}{}^1{}_{\sigma\mu}F_{DE}{}^{2\sigma\mu}\}$$

$$\mathcal{L}_{DSU(2)} = \text{Tr}\ \sqrt{g}[aF_W{}^1{}_{\sigma\mu}F_W{}^{2\sigma\mu}]$$

$$\mathcal{L}_{SU(3)} = \text{Tr}\ \sqrt{g}\{M[\partial_\nu + i(A_{SU(3)}{}^1{}_\nu + A_{SU(3)}{}^2{}_\nu)]F_{SU(3)}{}^1{}_{\sigma\mu}[\partial^\nu + i(A_{SU(3)}{}^{1\nu} + + A_{SU(3)}{}^{2\nu})]F_{SU(3)}{}^{2\sigma\mu} + aF_{SU(3)}{}^1{}_{\sigma\mu}F_{SU(3)}{}^{2\sigma\mu} - dA_{SU(3)}{}^2{}_\mu A_{SU(3)}{}^{2\mu}\}$$

---

84 The constants have the dimensions: M has the dimension of inverse mass squared, b has dimension mass squared, a is dimensionless, c and c' have dimension mass, and d has dimension mass squared.

85 We show gravity weakens at very short distances using our Two-Tier Quantum Field Theory formalism. See Appendix A, and Blaha (2003) and (2005a) among other books by the author.

86 We only consider the gauge field lagrangian terms.

$$\mathfrak{L}_{DSU(3)} = \text{Tr}\,\sqrt{g}\{M[\partial_\nu + i(A_{DSU(3)}{}^1{}_\nu + A_{DSU(3)}{}^2{}_\nu)]F_{DSU(3)}{}^1{}_{\sigma\mu}[\partial^\nu + i(A_{DSU(3)}{}^{1\nu} + A_{DSU(3)}{}^{2\nu})]F_{DSU(3)}{}^{2\sigma\mu} + aF_{DSU(3)}{}^1{}_{\sigma\mu}F_{DSU(3)}{}^{2\sigma\mu} - dA_{DSU(3)}{}^2{}_\mu A_{DSU(3)}{}^{2\mu}\}$$

$$\mathfrak{L}_U = \text{Tr}\,\sqrt{g}[aF_U{}^1{}_{\sigma\mu}F_U{}^{2\sigma\mu}]$$

$$\mathfrak{L}_V = \text{Tr}\,\sqrt{g}[aF_V{}^1{}_{\sigma\mu}F_V{}^{2\sigma\mu}]$$

$$\mathfrak{L}_{DU} = \text{Tr}\,\sqrt{g}[aF_{DU}{}^1{}_{\sigma\mu}F_{DU}{}^{2\sigma\mu}]$$

$$\mathfrak{L}_{DV} = \text{Tr}\,\sqrt{g}[aF_{DV}{}^1{}_{\sigma\mu}F_{DV}{}^{2\sigma\mu}]$$

$$\mathfrak{L}_S = \text{Tr}\,\sqrt{g}[aF_S{}^1{}_{\sigma\mu}F_S{}^{2\sigma\mu}]$$

$$\mathfrak{L}_G = \text{Tr}\,\sqrt{g}[MD_\nu R^1{}_{\sigma\mu}D^\nu R^{2\sigma\mu} + aR^1{}_{\sigma\mu}R^{2\sigma\mu} + bg^{\sigma\mu}(R^{1\beta}{}_{\sigma\beta\mu} + R^{2\beta}{}_{\sigma\beta\mu}) + cg^{\sigma\mu}g^2{}_{\sigma\mu} + c'g^{2\sigma\mu}g^2{}_{\sigma\mu}]$$
$$= \text{Tr}\,\sqrt{g}[MD_\nu R^1{}_{\sigma\mu}D^\nu R^{2\sigma\mu} + aR^1{}_{\sigma\mu}R^{2\sigma\mu} + bH + cg^{\sigma\mu}g^2{}_{\sigma\mu} + c'g^{2\sigma\mu}g^2{}_{\sigma\mu}]$$

$$\mathfrak{L}_{int} = \mathfrak{L} - (\mathfrak{L}_E + \mathfrak{L}_{SU(2)} + \mathfrak{L}_{DE} + \mathfrak{L}_{DSU(2)} + \mathfrak{L}_{SU(3)} + \mathfrak{L}_U + \mathfrak{L}_V + \mathfrak{L}_Y + \mathfrak{L}_{DSU(3)} + \mathfrak{L}_{DU} + \mathfrak{L}_{DV} + \mathfrak{L}_{DY} + \mathfrak{L}_S + \mathfrak{L}_G) \quad (5.26)$$

with appropriate sums over layers and gravitational B spinor connection terms omitted. Thus $\mathfrak{L}_{SU(3)}$, $\mathfrak{L}_{SU(2)}$, $\mathfrak{L}_E$, $\mathfrak{L}_{DE}$, $\mathfrak{L}_{DSU(2)}$, $\mathfrak{L}_U$,, $\mathfrak{L}_V$, $\mathfrak{L}_Y$, $\mathfrak{L}_S$, and parts of $\mathfrak{L}_{int}$ are the dominant interactions within hadrons, and $\mathfrak{L}_G$, $\mathfrak{L}_E$ and parts of $\mathfrak{L}_{int}$ are the dominant interactions in space within the framework of this discussion.

The $D_\nu R^1{}_{\sigma\mu}$ and $D^\nu R^{2\sigma\mu}$ terms have the form:

$$D_\nu R^i{}_{\sigma\mu} = + \partial_\nu R^i{}_{\sigma\mu} - H^{1\beta}{}_{\sigma\nu}R^i{}_{\beta\mu} - H^{1\beta}{}_{\nu\mu}R^i{}_{\sigma\beta} \quad (5.27)$$

for i = 1, 2 while covariant derivatives for internal symmetries are given by eq. 5.5a.

Blaha (2019g) and (2018e) discusses further details of the lagrangian obtained from the Riemann-Christoffel tensor, and its gravitation and Strong Interaction terms. The Dark Strong interaction terms generated by the first term in eq. 5.23 has an effect that parallels the Strong interaction case and leads to the confinement of Dark quarks.[87]

5.4 New Vector Boson Interactions

The above lagrangian can be broken up into pieces in the following manner:

$$\mathfrak{L}_E = \text{Tr}\,\sqrt{g}\{M\{[\partial_\nu + i(A_E{}^1{}_\nu + A_E{}^2{}_\nu)]F^1{}_{E\sigma\mu}[\partial^\nu + i(A_E{}^{1\nu} + A_E{}^{2\nu})]F^2{}_E{}^{\sigma\mu}\} + aF_E{}^1{}_{\sigma\mu}F_E{}^{2\sigma\mu}\} \quad (5.28)$$

$$\mathfrak{L}_{SU(2)} = \text{Tr}\,\sqrt{g}[aF_W{}^1{}_{\sigma\mu}F_W{}^{2\sigma\mu}]$$
$$\mathfrak{L}_{DE} = \text{Tr}\,\sqrt{g}\{M\{[\partial_\nu + i(A_{DE}{}^1{}_\nu + A_{DE}{}^2{}_\nu)]F^1{}_{DE\sigma\mu}[\partial^\nu + i(A_{DE}{}^{1\nu} + A_{DE}{}^{2\nu})]F_{DE}{}^{2\sigma\mu}\} +$$

[87] See Blaha (2019g) and (2018e).

$$+ aF_{DE}{}^{1}{}_{\sigma\mu}F_{DE}{}^{2\sigma\mu}\}$$

$$\mathcal{L}_{DSU(2)} = \text{Tr}\ \sqrt{g}[aF_W{}^{1}{}_{\sigma\mu}F_W{}^{2\sigma\mu}]$$

$$\mathcal{L}_{SU(3)} = \text{Tr}\ \sqrt{g}\{M[\partial_\nu + i(A_{SU(3)}{}^{1}{}_{\nu} + A_{SU(3)}{}^{2}{}_{\nu})]F_{SU(3)}{}^{1}{}_{\sigma\mu}[\partial^\nu + i(A_{SU(3)}{}^{1\nu} + A_{SU(3)}{}^{2\nu})]F_{SU(3)}{}^{2\sigma\mu} + aF_{SU(3)}{}^{1}{}_{\sigma\mu}F_{SU(3)}{}^{2\sigma\mu} - dA_{SU(3)}{}^{2}{}_{\mu}A_{SU(3)}{}^{2\mu}\} \tag{5.29}$$

$$\mathcal{L}_U = \text{Tr}\ \sqrt{g}[aF_U{}^{1}{}_{\sigma\mu}F_U{}^{2\sigma\mu}]$$

$$\mathcal{L}_V = \text{Tr}\ \sqrt{g}[aF_V{}^{1}{}_{\sigma\mu}F_V{}^{2\sigma\mu}]$$

$$\mathcal{L}_Y = \text{Tr}\ \sqrt{g}\{M\{[\partial_\nu + i(Y^1{}_\nu + Y^2{}_\nu)]F^1{}_{Y\sigma\mu}[\partial^\nu + i(Y^{1\nu} + Y^{2\nu})]F^2{}_Y{}^{\sigma\mu}\} + aF_Y{}^{1}{}_{\sigma\mu}F_Y{}^{2\sigma\mu}\}$$

$$\mathcal{L}_{DY} = \text{Tr}\ \sqrt{g}\{M\{[\partial_\nu + i(Y_D{}^1{}_\nu + Y_D{}^2{}_\nu)]F^1{}_{DY\sigma\mu}[\partial^\nu + i(Y_D{}^{1\nu} + Y_D{}^{2\nu})]F_{DY}{}^{2\sigma\mu}\} + aF_{DY}{}^{1}{}_{\sigma\mu}F_{DY}{}^{2\sigma\mu}\}$$

$$\mathcal{L}_S = \text{Tr}\ \sqrt{g}[aF_S{}^{1}{}_{\sigma\mu}F_S{}^{2\sigma\mu}]$$

$$\mathcal{L}_G = \text{Tr}\ \sqrt{g}[MD_\nu R^1{}_{\sigma\mu}D^\nu R^{2\sigma\mu} + aR^1{}_{\sigma\mu}R^{2\sigma\mu} + bg^{\sigma\mu}(R^{1\beta}{}_{\sigma\beta\mu} + R^{2\beta}{}_{\sigma\beta\mu}) + cg^{\sigma\mu}g^2{}_{\sigma\mu} + c'g^{2\sigma\mu}g^2{}_{\sigma\mu}]$$

$$= \text{Tr}\ \sqrt{g}[MD_\nu R^1{}_{\sigma\mu}D^\nu R^{2\sigma\mu} + aR^1{}_{\sigma\mu}R^{2\sigma\mu} + bH + cg^{\sigma\mu}g^2{}_{\sigma\mu} + c'g^{2\sigma\mu}g^2{}_{\sigma\mu}]$$

$$\mathcal{L}_{int} = \mathcal{L} - (\mathcal{L}_E + \mathcal{L}_{SU(2)} + \mathcal{L}_{DE} + \mathcal{L}_{DSU(2)} + \mathcal{L}_{SU(3)} + \mathcal{L}_U + \mathcal{L}_V + \mathcal{L}_Y + \mathcal{L}_S + \mathcal{L}_G) \tag{5.30}$$

again with appropriate sums over layers and with coupling constants not displayed to avoid clutter. Thus $\mathcal{L}_{SU(3)}$, $\mathcal{L}_{SU(2)}$, $\mathcal{L}_E$, $\mathcal{L}_{DE}$, $\mathcal{L}_{DSU(2)}$, $\mathcal{L}_U$,, $\mathcal{L}_V$, $\mathcal{L}_Y$, $\mathcal{L}_S$, and parts of $\mathcal{L}_{int}$ are the dominant interactions within hadrons, and $\mathcal{L}_G$, $\mathcal{L}_E$ and parts of $\mathcal{L}_{int}$ are the dominant interactions in space within the framework of this discussion. The terms of $\mathcal{L}_{int}$ have 'new' interactions between gauge fields that are described in some detail in Blaha (2017b) and other books. These interactions are not in the conventional Standard Model. They lead to modifications of gravity, the Strong Interactions, spin dynamics and so on.

6. Other QUeST-UST Topics

There are a number of important UST topics that are presented in Blaha (2020c) and (2018e) as well as other book by the author. They include:

1. Color Confinement
2. Gravitation Potential at Earthly, Galactic, and intergalactic Distances
3. Higgs Mechanism and Symmetry Breaking. Fermion and Gauge Boson Masses
4. Monads From factorization of Wave Functions to Eliminate Quantum Entanglement
5. Determination of α and Other Coupling Constants
6. Evidence for Faster-than-Light Particles and their Implications
7. Complex Gravitation and the Species group
8. An Equipartition Principle for Universe Matter and Energy Abundance

QUeST-UST is a complete theory of the framework of elementary particle phenomena.

7. Origin of Megaverse

This chapter describes the possible birth of the 1024 dimension UTMOST Megaverse from a fermion with one internal dimension residing in a separate 10 dimension space.[88] It describes the BMOST theory developed in 2020 in earlier books by the author.

7.1 BMOST Origin

BMOST[89] assumes an initial one dimension space and one fermion. The fermion is introduced to implement the fermion-dimension duality found in UTMOST and QUeST.

The derivation of the 32 × 32 = 1024 dimension UTMOST array requires a fermion, which we call the *urfermion*,[90] to be an 10 dimension fermion. The 32 spinor components of the 10 dimension urfermion will be used to generate the 32 × 32 dimension array of UTMOST.

We propose a picture of a 10 dimension fermion —the urfermion—that acts as the seed of the subsequent Megaverse. The evolution (perhaps instantaneous) of the Megaverse begins with a seed of great energy, and becomes the 1024 dimension UTMOST Megaverse. It then generates internal symmetries and the space-time of UTMOST with symmetry breakdown. Subsequently universes appear within the Megaverse as described in chapter 1.

7.2 Dynamics of the URfermion

The urfermion must have a dynamics that enables it to generate the 1024 dimension array. Since the target array is not symmetric the PseudoQuantum formulation[91] of Quantum Field Theory will be seen to be required. The derivation of the "birth" of the Megaverse parallels that of BQUeST in chapter 1.

We define an electromagnetic-like model lagrangian with the urfermion represented by *two* quantum fields[92] ψ_1 and ψ_2:

$$\mathscr{L} = F^{1\mu\nu} F^2{}_{\mu\nu} + \overline{\psi}_2\gamma^0(i\gamma\cdot\partial - m)\psi_1 + \overline{\psi}_1\gamma^0(i\gamma\cdot\partial - m)\psi_2 - e_0\overline{\psi}_2\gamma^0\gamma\cdot A_2\psi_1 - \overline{\psi}_1\gamma^0\gamma\cdot A_1\psi_2 \tag{7.1}$$

where m is the mass-energy of the produced Megaverse, and

$$F^i{}_{\mu\nu} = \partial_\nu A_{i\mu} - \partial_\mu A_{i\nu} \tag{7.2}$$

---

[88] The prevalence of 10 dimension spaces in SuperString theories raises the possibility of a relation of the Megaverse to a SuperString space.

[89] See Blaha (2020d) and later books.

[90] We use the Germanic prefix ur- to signify original (or earliest). The author used this prefix in the 1970s in published papers on quarks and leptons including discrete scaling fermion masses.

[91] S. Blaha, Il Nuovo Cimento **49A**, 35 (1979). Reproduced as Appendix 1-B below.

[92] We use a PseudoQuantum Electromagnetic-like pair of fields also. The electromagnetic-like particle is a model of the universe. See eqs. 61 – 90 in Appendix 1-A.

The Megaverse particle is represented by two fields: A_1^{μ} and A_2^{μ}. The dynamical equations in the Lorentz gauge are

$$\Box A_{i\mu} = e_0 J_{i\mu} \tag{7.3}$$

where

$$J_{1\mu} = \bar{\psi}_2 \gamma^0 \gamma_\mu \psi_1 \tag{7.4}$$

$$J_{2\mu} = \bar{\psi}_1 \gamma^0 \gamma_\mu \psi_2 \tag{7.5}$$

The currents have the conservation laws:

$$\partial^\mu J_{i\mu} = 0$$

7.3 Indices and Dimensions

Eqs. 7.2 – 7.5 hold for the 10 dimension urfermion. If we take the urfermion off-shell then we can separate eqs. 7.3 – 7.5 into their respective spinor components:

$$\Box A_{i\mu}{}^{ab} = e_0 J_{i\mu}{}^{ab} \tag{7.6}$$

where

$$J_{1\mu}{}^{ab} = \bar{\psi}_1{}^a \gamma^0 \gamma_\mu \psi_2{}^b \tag{7.7}$$

$$J_{2\mu}{}^{ab} = \bar{\psi}_2{}^a \gamma^0 \gamma_\mu \psi_1{}^b \tag{7.8}$$

The (conserved) charge densities ($\mu = 0$) are

$$J_{10}{}^{ab} = \bar{\psi}_1{}^a \psi_2{}^b \tag{7.9}$$

$$J_{20}{}^{ab} = \bar{\psi}_2{}^a \psi_1{}^b \tag{7.10}$$

The independence of the spinor components of each of the two urfermion fields guarantees an array of independent indices, and thence an array of independent dimensions. .

7.3.1 The Difference between Indices and Dimensions

Clearly $A_{i0}{}^{ab}$ is a 32 × 32 array since the urfermion exists in a 10 dimension space and has 32 independent component spinors. The array is not symmetric because we use a PseudoQuantum framework.

In section 1.3.1 we showed that indices can be taken to be the internal dimensions within an entity.

We thus can conclude the indices of $A_{i\mu}{}^{ab}$ can be viewed as specifying coordinates *and dimensions* within $A_{i\mu}$. We have mapped the urfermion to a 32 × 32 = 1024 dimension array. This array serves as the dimensions of the UTMOST space.

7.4 Origin of Megaverse

We can then envision the possibility that the Megaverse started as 1-dimensional urfermion with one internal dimension, and then "acquired" the 1024 dimensions and 1024 fundamental fermions of UTMOST. This process could proceed, as it likely does, instantaneously. A corollary benefit is the location of the urfermion in a 10 dimension external space.

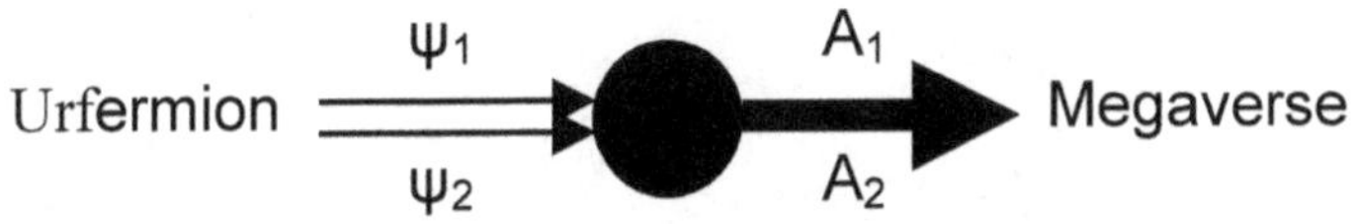

Figure 7.1. Diagram for the transition of the urfermion to a Megaverse.

We conclude the1024 dimension UTMOST universe arises from one fermion residing in an 10-dimension external space. The urfermion has one internal dimension that grows to the 1024 dimensions of the UTMOST universe.

Appendix 7-A. Features of UTMOST Octonion Space

This appendix describes the 64 complex octonion dimension space of UTMOST. It also discusses an alternate 32 *quaternion octonion*[93] dimension space for UTMOST. Given the present state of physical data these variants are both acceptable. and features of QUeST.

7-A.1 UTMOST Space

An octonion contains eight dimensions. A complex octonion contains sixteen dimensions. A *quaternion octonion*[94] contains 32 dimensions. Fig. 7-A.1 depicts the 64 dimension complex octonion space as a 64 × 16 array of dimensions. It uses a "dot" or pebble • to represent a dimension[95]. The dimensions of the space are not assigned physically until they are mapped to internal symmetry group fundamental representation dimensions and space-time dimensions. Rather than create a cumbersome coordinate-based notation we choose to use •'s.

Figure 7-A.1. The 64 complex octonion dimension UTMOST array. This is the 64 × 16 array of •'s. It has 1024 dimensions.

The UTMOST space can also be viewed as a space of 32 quaternion octonion dimensions. It also has a total of 1024 dimensions. Fig. 7-A.2 creates a 32 × 32 array of dimensions for this space. *A 32 × 32 array is important for the derivation of UTMOST from one-dimension BMOST.*

The 32 × 32 form of the UTMOST dimension array is based on a 32 quaternion octonion dimension space. The difference between this form of space and the 64 complex octonion dimension space above is not physically meaningful at present. The

[93] A quaternion octonion is a 32 dimension set of coordinates. It is the composition of a quaternion and an octonion that could be represented by an expression such as q(o), which represents an octonion argument o for a quaternion functional q that expands each dimension in the octonion fourfold It is a generalization of a complex octonion..
[94] This notation, originated here, is for the composition of a quaternion and an octonion.
[95] The use of pebbles is a useful simplification of coordinates that enables assignments of pebbles (coordinates) to group representations to be visual.

difference will be physically meaningful if the masses of the fermion spectrum and the full pattern of symmetry breaking are determined. Then one can differentiate between the symmetry group spectrum and mass spectrums of the respective possibilities.

•••••••• •••••••• •••••••• ••••••••
•••••••• •••••••• •••••••• ••••••••
•••••••• •••••••• •••••••• ••••••••
•••••••• •••••••• •••••••• ••••••••
•••••••• •••••••• •••••••• ••••••••
•••
•••••••• •••••••• •••••••• ••••••••

Figure 7-A.2. The UTMOST array with 32 × 32 dimensions for a 32 quaternion octonion dimension space.

The repetitive pattern of groups seen in QUeST leads us to assume that UTMOST has a similar repetitive pattern. We will use a four layer format for the 32 × 32 array of dimensions. Each layer consists of 8 rows of Fig. 7-A.2. Each layer can be put in a form analogous to Fig. 1-A.4 (and to Fig. 1-A.9). See Fig. 7-A.3.

We map between dimensions and fundamental group representations. We use the maps in Table 7-A.1 to set up the group ↔ dimension map, bearing in mind the group representations of the Standard Model:

U(4)	↔	8 real dimensions
U(2)	↔	4 real dimensions
SU(3)	↔	6 real dimensions
U(1)⊗SU(2)	↔	4 real dimensions

Table 7-A.1. Map between fundamental representations and their dimensions.

Fig. 7-A.3 shows the content of one UTMOST layer. The four layers of UTMOST are four copies of Fig. 7-A.3.[96] The separation of the set of dimensions is accomplished by following the procedure given earlier.[97]

Fig. 7-A.5 shows the four layers (each in two rows) of the 32 × 32 dimension UTMOST array, which is composed of 4 × 4 blocks. The 4 × 4 blocks are within the

[96] The Layer groups are U(4) groups. They mix the generations of each of the top four layers, generation by generation, separately from the Layer groups mixing the lower four layers. This feature enables QUeST universes to be generated from either the top four layers or the lower four layers.

[97] The separation of the dimensions into the subgroup factors' representation can be implemented as group transformations and definitions using standard group theoretic methods. A more formal method for extracting the subgroup content of representations uses a symmetric group analysis of U(n) representation characters. See S. Blaha, J. Math. Phys. **10**, 2156 (1969) for a detailed discussion of this approach.

four block 8 × 8 sections for each pair: Normal+Dark1, Dark2+Dark3, Dark4+Dark5 and Dark6+Dark7. In total they form the 32 × 32 = 1024 UTMOST dimension array.

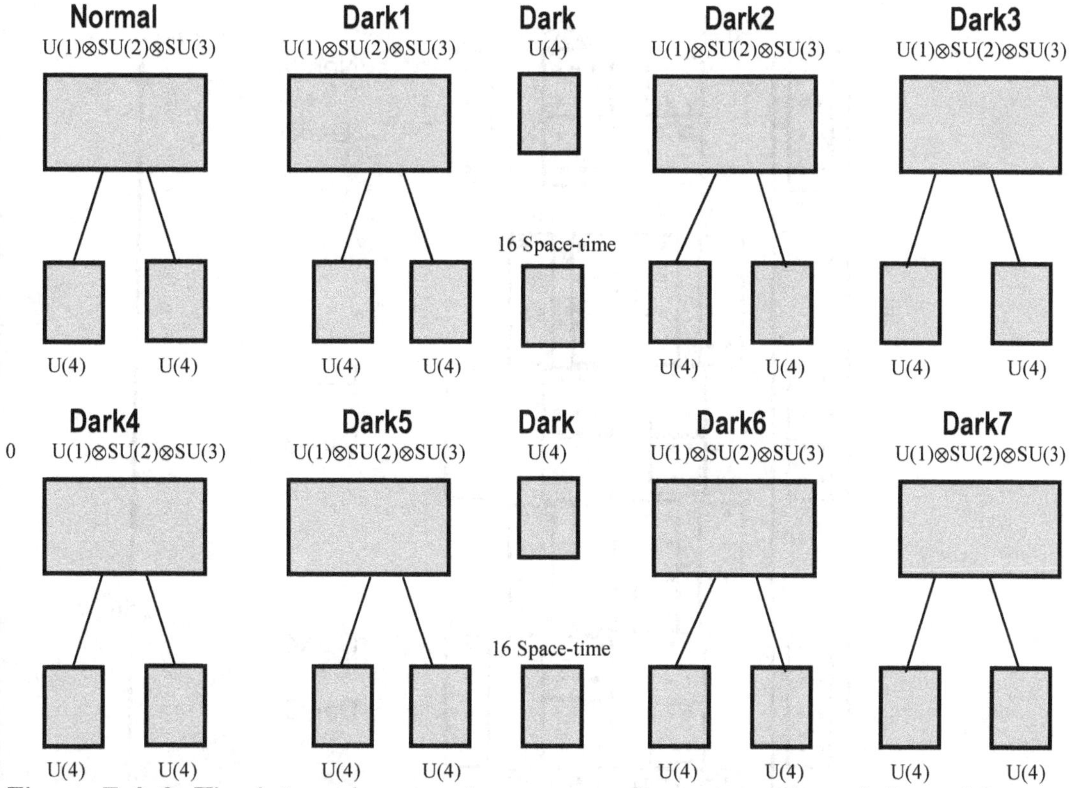

Figure 7-A.3. The internal symmetry groups of *one layer* (consisting of 8 rows in Fig. 7-A.2) of the four layers of 32 × 32 dimension UTMOST. The other three layers are copies of the this layer. Note the Dark U(4) groups. One U(4) "rotates" among Normal, Dark1, Dark2, and Dark3. The other U(4) "rotates" among Dark4, Dark5, Dark6, and Dark7.

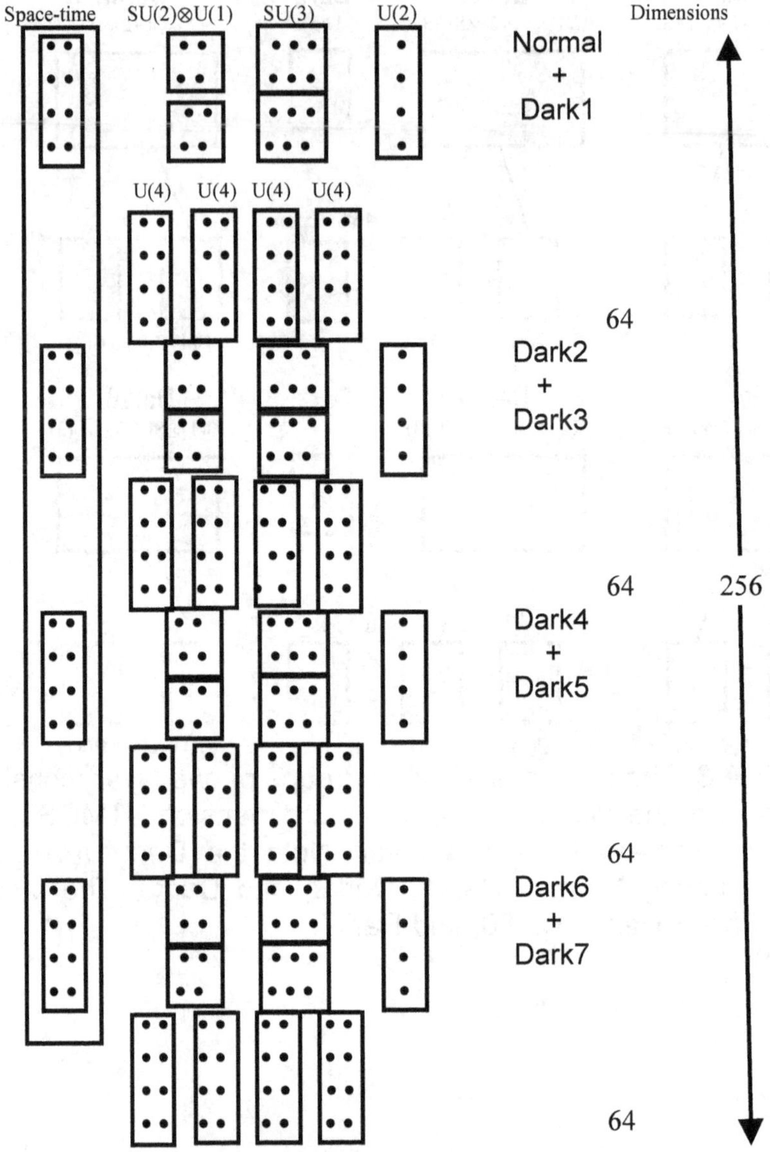

Figure 7-A.4 The *first* of the four layers of UTMOST dimensions with boxes around sets of dimensions for fundamental group representations. The U(4) Dark groups have been separated into U(2)⊗U(2) factors for later use.

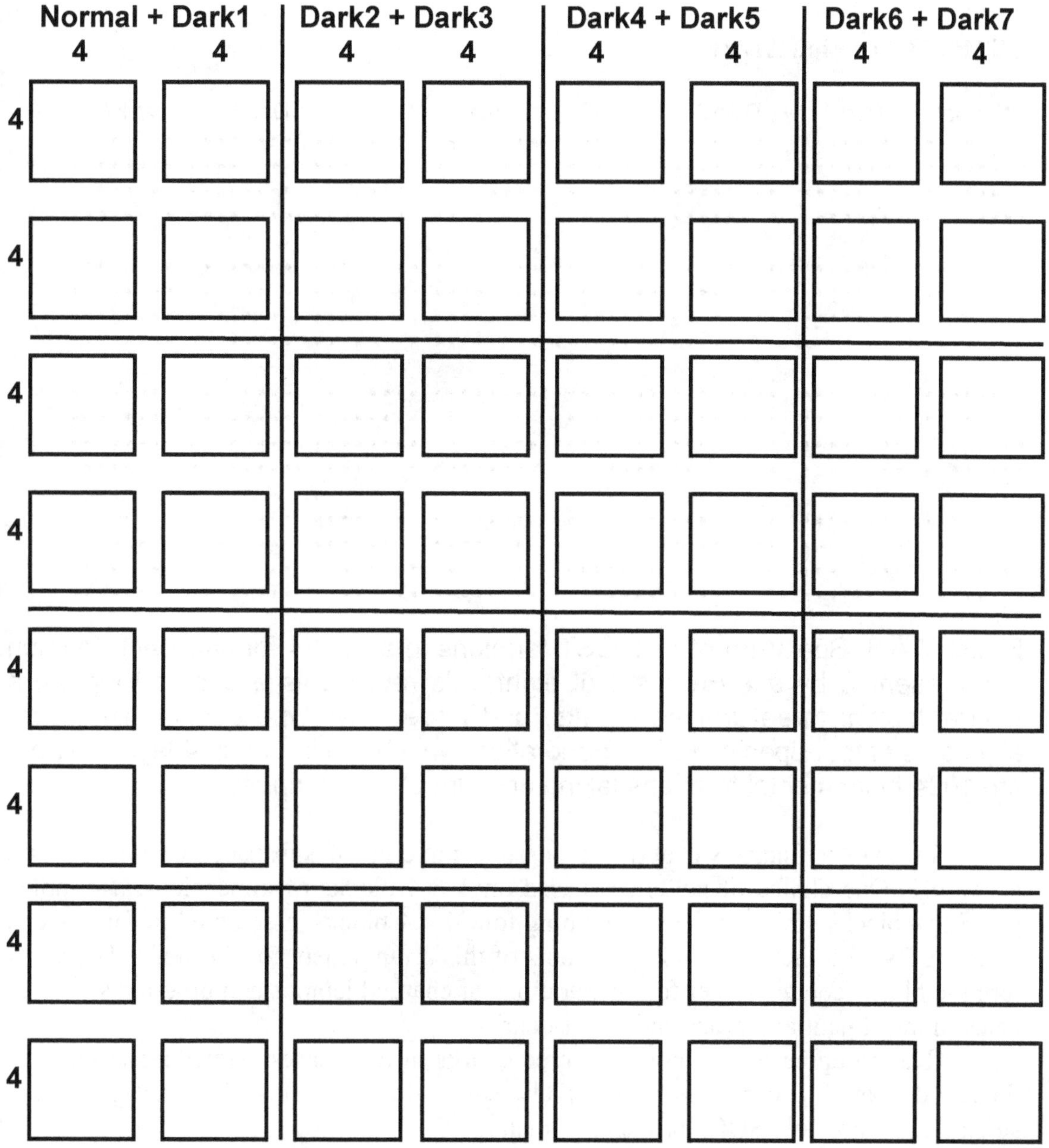

Figure 7-A.5. Four layers (each in two rows) in the 32 × 32 dimension UTMOST array composed of 4 × 4 blocks, which are within the four block 8 × 8 sections for each pair: Normal+Dark1, Dark2+Dark3, Dark4+Dark5 and Dark6+Dark7. In total they form the 32 × 32 = 1024 UTMOST dimension array.

7-A.2 UTMOST Fermions

Given the form of the internal symmetries in UTMOST we can determine the fermions in the fundamental group representations as shown in Fig. 7-A.6.

UTMOST Fermion Array

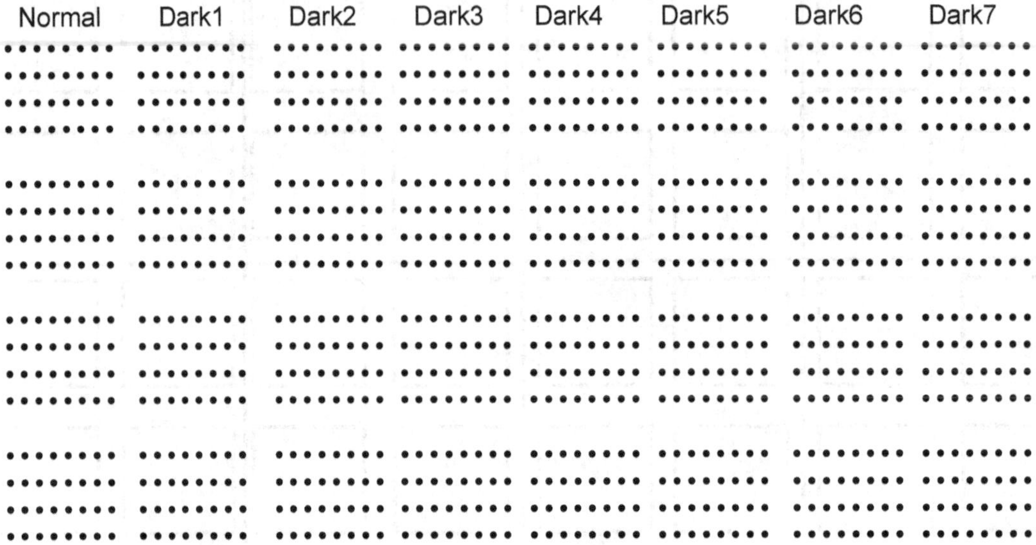

Figure 7-A.6. Spectrum of UTMOST fermions in a 16×64 format. Each fermion is represented by a •..Each set of eight •.'s represents a charged lepton, a neutral lepton, three up-type quarks, and three down-type quarks. There are eight sets of four species in four generations which are in turn in 4 layers. There are 1024 fundamental fermions taking account of quark triplets.

In chapter 3qqzz we outlined possible patterns of subspaces of QUeST and UTMOST. One choice of pattern is based on 4 × 4 blocks of dimensions, assembled into 8 × 8 blocks of dimensions containing four 4 × 4 blocks, assembled in four layers. Fig. 7-A.7 shows the possible implications of this arrangement for fermions. The 4 × 4 fermion blocks contain either four generations of charged leptons and up-quarks, or four generations of neutral leptons and down-quarks.

The grouping of a lepton and three quarks in both cases creates a similarity to time and spatial coordinates respectively suggesting a broken Lorentz group-like structure or a possible SU(4) broken symmetry.

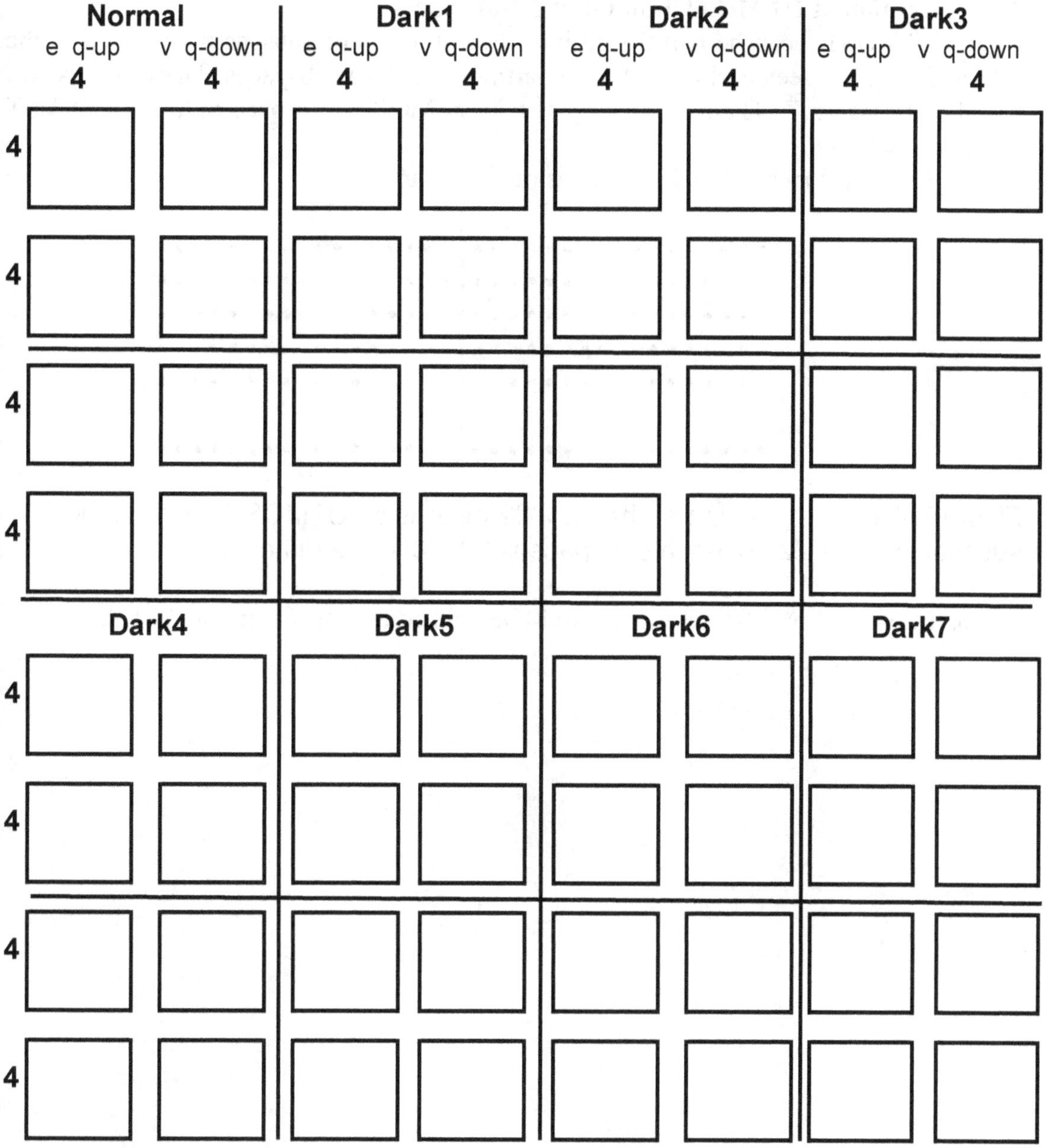

Figure 7-A.7. Block form of the 32 × 32 UTMOST fermion array with each row corresponding to *half of an UTMOST layer*. Thus 8 × ½ = 4 layers results. Each block contains four generations of fermions. The result is sixty-four 4 × 4 blocks. The label e q-up indicates a charged lepton – up-type quark pair, v q-down indicates a neutral lepton – down-type quark pair, and so on. *The form displayed here may explain why generations come in fours.*

7-A.3 Partition of UTMOST into QUeST Subspaces

UTMOST[98] can be partitioned into several levels of subspaces. We modify the UTMOST figures seen earlier to show a partition into two subspaces. These spaces will be MOST spaces.[99] Then we can partition a MOST subspace into two QUeST subspaces and so on.

The partitioned UTMOST dimension arrays are:

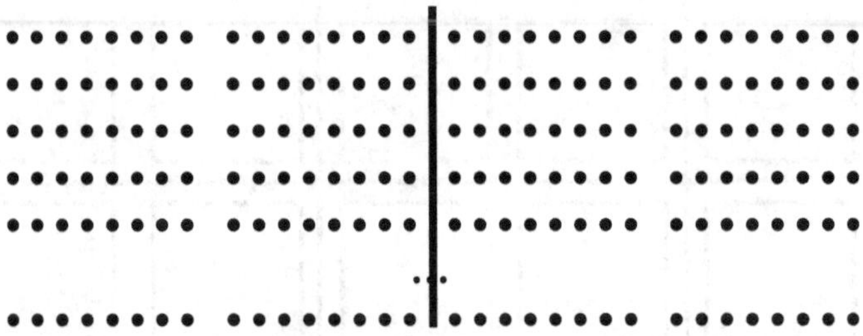

Figure 7-A.8. The partition of the 32 × 32 dimension UTMOST array into MOST subspaces The size of each subspace is 32×16 = 512 dimensions.

The partition of UTMOST fermions into MOST fermions appears in Fig. 7-A.9.

[98] We will consider MOST subspaces below.
[99] MOST is described in Blaha (2020i) and earlier books by the author.

UTMOST Fermion Array Partitioned into MOST Fermion Arrays

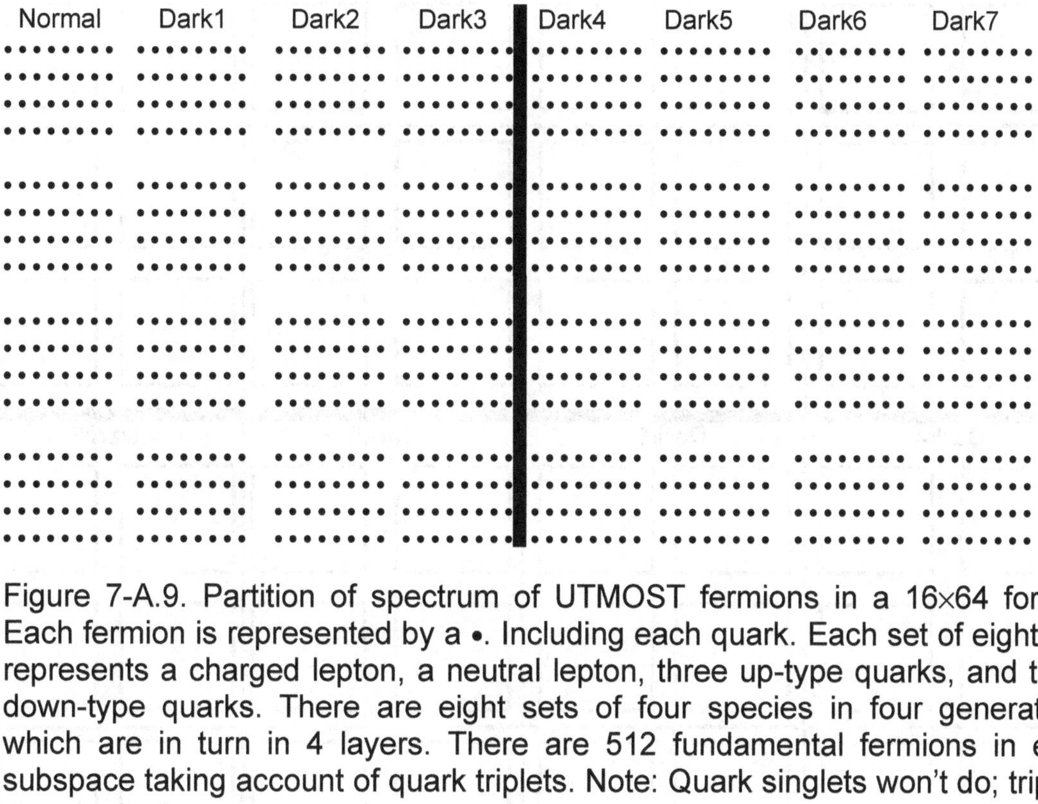

Figure 7-A.9. Partition of spectrum of UTMOST fermions in a 16×64 format. Each fermion is represented by a •. Including each quark. Each set of eight •.'s represents a charged lepton, a neutral lepton, three up-type quarks, and three down-type quarks. There are eight sets of four species in four generations which are in turn in 4 layers. There are 512 fundamental fermions in each subspace taking account of quark triplets. Note: Quark singlets won't do; triplets are required.

UTMOST Fermion Array Partitioned into MOST Fermion Arrays with 4 × 4 Blocks

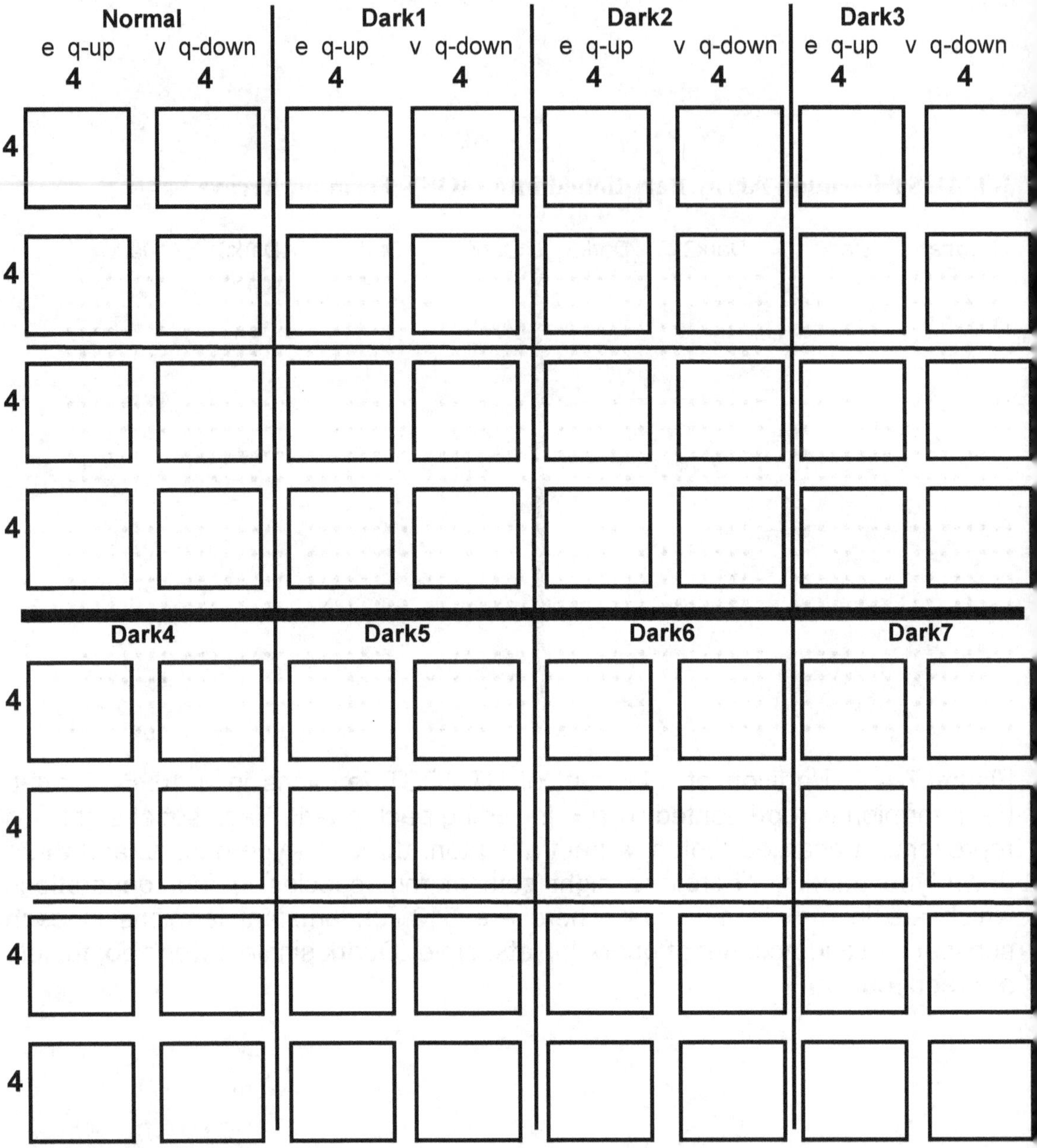

Figure 7-A.10. Partition of block form of the 32 × 32 UTMOST fermion array with each row corresponding to *half of an UTMOST layer*. Thus 8 × ½ = 4 layers results. Each block contains four generations of fermions. The result is sixty-four 4 × 4 blocks. The label e q-up indicates a charged lepton – up-type quark pair, v q-down indicates a neutral lepton – down-type quark pair, and so on.

8. The Connection of QUeST and UTMOST

Universes are embedded within the Megaverse. A QUeST universe has four complex quaternion space-time dimensions. The UTMOST Megaverse has eight complex quaternion space-time dimensions. We can therefore embedded QUeST universes within the Megaverse. In fact, any number of QUeST universes can be embedded in the UTMOST Megaverse.

At a deeper level we found we could define a basis for QUeST, called BQUeST, having one *internal* dimension and consisting of one seed fermion residing in an eight dimension space-time (Chapter 1). The seed fermion evolves into a universe. We can choose the space of the fermion to be the UTMOST Megaverse since it has an eight complex quaternion space-time.[100] The fermion (universe) develops the *internal* 256 dimension space of QUeST (perhaps instantaneously). Thus we obtain a QUeST universe within the UTMOST Megaverse.

The Megaverse itself also has a possible origin in a one dimension – one urfermion state within BMOST. The urfermion evolves into the UTMOST Megaverse. The fermion must reside in a ten dimension[101] space in order for UTMOST's 1024 dimensions to emerge (Chapter 7). This space may be related to a yet deeper SuperString theory. Its details remain to be disclosed.

The BMOST fermion develops the 1024 internal dimensions of the UTMOST Megaverse.

As a result we have a connection of our universe to the Megaverse, which in turn may be connected to an unknown ten dimension space.

[100] As we saw in section 1-A.3 we can partition the universe and Megaverse to real coordinates.
[101] The details of the space remain to be found. Although it may be quaternionic or octonionic we can assume that it has a real coordinates partition for the present.

REFERENCES

Akhiezer, N. I., Frink, A. H. (tr), 1962, *The Calculus of Variations* (Blaisdell Publishing, New York, 1962).

Bjorken, J. D., Drell, S. D., 1964, *Relativistic Quantum Mechanics* (McGraw-Hill, New York, 1965).

Bjorken, J. D., Drell, S. D., 1965, *Relativistic Quantum Fields* (McGraw-Hill, New York, 1965).

Blaha, S., 1998, *Cosmos and Consciousness* (Pingree-Hill Publishing, Auburn, NH, 1998).

_______, 2002, *A Finite Unified Quantum Field Theory of the Elementary Particle Standard Model and Quantum Gravity Based on New Quantum Dimensions™ & a New Paradigm in the Calculus of Variations* (Pingree-Hill Publishing, Auburn, NH, 2002).

_______, 2003, *A Finite Unified Quantum Field Theory of the Elementary Particle Standard Model and Quantum Gravity Based on New Quantum Dimensions™ and a New Paradigm in the Calculus of Variations* (Pingree-Hill Publishing, Auburn, NH, 2003).

_______, 2004, *Quantum Big Bang Cosmology: Complex Space-time General Relativity, Quantum Coordinates™ Dodecahedral Universe, Inflation, and New Spin 0, ½, 1 & 2 Tachyons & Imagyons* (Pingree-Hill Publishing, Auburn, NH, 2004).

______, *2005a, Quantum Theory of the Third Kind: A New Type of Divergence-free Quantum Field Theory Supporting a Unified Standard Model of Elementary Particles and Quantum Gravity based on a New Method in the Calculus of Variations* (Pingree-Hill Publishing, Auburn, NH, 2005).

______, 2005b, *The Metatheory of Physics Theories, and the Theory of Everything as a Quantum Computer Language* (Pingree-Hill Publishing, Auburn, NH, 2005).

______, 2005c, *The Equivalence of Elementary Particle Theories and Computer Languages: Quantum Computers, Turing Machines, Standard Model, Superstring Theory, and a Proof that Gödel's Theorem Implies Nature Must Be Quantum* (Pingree-Hill Publishing, Auburn, NH, 2005).

______, 2006a, *The Foundation of the Forces of Nature* (Pingree-Hill Publishing, Auburn, NH, 2006).

______, 2006b, *A Derivation of ElectroWeak Theory based on an Extension of Special Relativity; Black Hole Tachyons; & Tachyons of Any Spin.* (Pingree-Hill Publishing, Auburn, NH, 2006).

______, 2007a, *Physics Beyond the Light Barrier: The Source of Parity Violation, Tachyons, and A Derivation of Standard Model Features* (Pingree-Hill Publishing, Auburn, NH, 2007).

______, 2007b, *The Origin of the Standard Model: The Genesis of Four Quark and Lepton Species, Parity Violation, the ElectroWeak Sector, Color SU(3), Three Visible Generations of Fermions, and One Generation of Dark Matter with Dark Energy* (Pingree-Hill Publishing, Auburn, NH, 2007).

______, *2008a, A Direct Derivation of the Form of the Standard Model From GL(16) (Pingree-Hill Publishing, Auburn, NH, 2008).*

______, 2008b, *A Complete Derivation of the Form of the Standard Model With a New Method to Generate Particle Masses Second Edition* (Pingree-Hill Publishing, Auburn, NH, 2008)

______, 2009, *The Algebra of Thought & Reality: The Mathematical Basis for Plato's Theory of Ideas, and Reality Extended to Include A Priori Observers and Space-Time Second Edition* (Pingree-Hill Publishing, Auburn, NH, 2009).

______, 2010a, *Operator Metaphysics: A New Metaphysics Based on a New Operator Logic and a New Quantum Operator Logic that Lead to a Mathematical Basis for Plato's Theory of Ideas and Reality* (Pingree-Hill Publishing, Auburn, NH, 2010).

______, 2010b, *The Standard Model's Form Derived from Operator Logic, Superluminal Transformations and GL(16)* (Pingree-Hill Publishing, Auburn, NH, 2010).

______, 2010c, *SuperCivilizations: Civilizations as Superorganisms* (McMann-Fisher Publishing, Auburn, NH, 2010).

______, 2011a, *21st Century Natural Philosophy Of Ultimate Physical Reality* (McMann-Fisher Publishing, Auburn, NH, 2011).

______, 2011b, *All the Universe! Faster Than Light Tachyon Quark Starships & Particle Accelerators with the LHC as a Prototype Starship Drive Scientific Edition* (Pingree-Hill Publishing, Auburn, NH, 2011).

______, 2011c, *From Asynchronous Logic to The Standard Model to Superflight to the Stars* (Blaha Research, Auburn, NH, 2011).

______, 2012a, *From Asynchronous Logic to The Standard Model to Superflight to the Stars volume 2: Superluminal CP and CPT, U(4) Complex General Relativity and The Standard Model, Complex Vierbein General Relativity, Kinetic Theory, Thermodynamics* (Blaha Research, Auburn, NH, 2012).

______, 2012b, *Standard Model Symmetries, And Four And Sixteen Dimension Complex Relativity; The Origin Of Higgs Mass Terms* (Blaha Reasearch, Auburn, NH, 2012).

______, 2013a, *Multi-Stage Space Guns, Micro-Pulse Nuclear Rockets, and Faster-Than-Light Quark-Gluon Ion Drive Starships* (Blaha Research, Auburn, NH, 2013).

______, 2013b, *The Bridge to Dark Matter; A New Sister Universe; Dark Energy; Inflatons; Quantum Big Bang; Superluminal Physics; An Extended Standard Model Based on Geometry* (Blaha Reasearch, Auburn, NH, 2013).

______, 2014a, *Universes and Megaverses: From a New Standard Model to a Physical Megaverse; The Big Bang; Our Sister Universe's Wormhole; Origin of the Cosmological Constant, Spatial Asymmetry of the Universe, and its Web of Galaxies; A Baryonic Field between Universes and Particles; Megaverse Extended Wheeler-DeWitt Equation* (Blaha Reasearch, Auburn, NH, 2014).

______, 2014b, *All the Megaverse! Starships Exploring the Endless Universes of the Cosmos Using the Baryonic Force* (Blaha Research, Auburn, NH, 2014).

______, 2014c, *All the Megaverse! II Between Megaverse Universes: Quantum Entanglement Explained by the Megaverse Coherent Baryonic Radiation Devices – PHASERs Neutron Star Megaverse Slingshot Dynamics Spiritual and UFO Events, and the Megaverse Microscopic Entry into the Megaverse* (Blaha Research, Auburn, NH, 2014).

______, 2015a, *PHYSICS IS LOGIC PAINTED ON THE VOID: Origin of Bare Masses and The Standard Model in Logic, U(4) Origin of the Generations, Normal and Dark Baryonic Forces, Dark Matter, Dark Energy, The Big Bang, Complex General Relativity, A Megaverse of Universe Particles* (Blaha Research, Auburn, NH, 2015).

______, 2015b, *PHYSICS IS LOGIC Part II: The Theory of Everything, The Megaverse Theory of Everything, U(4)⊗U(4) Grand Unified Theory (GUT), Inertial Mass = Gravitational Mass, Unified Extended Standard Model and a New Complex General Relativity with Higgs Particles, Generation Group Higgs Particles* (Blaha Research, Auburn, NH, 2015).

______, 2015c, *The Origin of Higgs ("God") Particles and the Higgs Mechanism: Physics is Logic III, Beyond Higgs – A Revamped Theory With a Local Arrow of Time, The Theory of Everything Enhanced, Why Inertial Frames are Special, Universes of the Mind* (Blaha Research, Auburn, NH, 2015).

______, 2015d, *The Origin of the Eight Coupling Constants of The Theory of Everything: U(8) Grand Unified Theory of Everything (GUTE), S^8 Coupling Constant Symmetry, Space-Time Dependent Coupling Constants, Big Bang Vacuum Coupling Constants, Physics is Logic IV* (Blaha Research, Auburn, NH, 2015).

______, 2016a, *New Types of Dark Matter, Big Bang Equipartition, and A New U(4) Symmetry in the Theory of Everything: Equipartition Principle for Fermions, Matter is 83.33% Dark, Penetrating the Veil of the Big Bang, Explicit QFT Quark Confinement and Charmonium, Physics is Logic V* (Blaha Research, Auburn, NH, 2016).

______, 2016b, *The Periodic Table of the 192 Quarks and Leptons in The Theory of Everything: The U(4) Layer Group, Physics is Logic VI* (Blaha Research, Auburn, NH, 2016).

______, 2016c, *New Boson Quantum Field Theory, Dark Matter Dynamics, Dark Matter Fermion Layer Mixing, Genesis of Higgs Particles, New Layer Higgs Masses, Higgs Coupling Constants, Non-Abelian Higgs Gauge Fields, Physics is Logic VII* (Blaha Research, Auburn, NH, 2016).

______, 2016d, *Unification of the Strong Interactions and Gravitation: Quark Confinement Linked to Modified Short-Distance Gravity; Physics is Logic VIII* (Blaha Research, Auburn, NH, 2016).

______, 2016e, *MoND: Unification of the Strong Interactions and Gravitation II, Quark Confinement Linked to Large-Scale Gravity, Physics is Logic IX* (Blaha Research, Auburn, NH, 2016).

______, 2016f, *CQ Mechanics: A Unification of Quantum & Classical Mechanics, Quantum/Semi-Classical Entanglement, Quantum/Classical Path Integrals, Quantum/Classical Chaos* (Blaha Research, Auburn, NH, 2016).

______, 2016g, *GEMS: Unified Gravity, ElectroMagnetic and Strong Interactions: Manifest Quark Confinement, A Solution for the Proton Spin Puzzle, Modified Gravity on the Galactic Scale* (Pingree Hill Publishing, Auburn, NH, 2016).

______, 2016h, *Unification of the Seven Boson Interactions based on the Riemann-Christoffel Curvature Tensor* (Pingree Hill Publishing, Auburn, NH, 2016).

______, 2017a, *Unification of the Eleven Boson Interactions based on 'Rotations of Interactions'* (Pingree Hill Publishing, Auburn, NH, 2017).

______, 2017b, *The Origin of Fermions and Bosons, and Their Unification* (Pingree Hill Publishing, Auburn, NH, 2017).

______, 2017c, *Megaverse: The Universe of Universes* (Pingree Hill Publishing, Auburn, NH, 2017).

______, 2017d, *SuperSymmetry and the Unified SuperStandard Model* (Pingree Hill Publishing, Auburn, NH, 2017).

______, 2017e, *From Qubits to the Unified SuperStandard Model with Embedded SuperStrings: A Derivation* (Pingree Hill Publishing, Auburn, NH, 2017).

______, 2017f, *The Unified SuperStandard Model in Our Universe and the Megaverse: Quarks, ...* , (Pingree Hill Publishing, Auburn, NH, 2017).

______, 2018a, *The Unified SuperStandard Model and the Megaverse SECOND EDITION A Deeper Theory based on a New Particle Functional Space that Explicates Quantum Entanglement Spookiness (Volume 1)* (Pingree Hill Publishing, Auburn, NH, 2018).

______, 2018b, *Cosmos Creation: The Unified SuperStandard Model, Volume 2, SECOND EDITION* (Pingree Hill Publishing, Auburn, NH, 2018).

______, 2018c, *God Theory (*Pingree Hill Publishing, Auburn, NH, 2018).

______, 2018d, *Immortal Eye: God Theory: Second Edition* (Pingree Hill Publishing, Auburn, NH, 2018).

______, 2018e, *Unification of God Theory and Unified SuperStandard Model THIRD EDITION* (Pingree Hill Publishing, Auburn, NH, 2018).

______, 2019a, *Calculation of: QED α = 1/137, and Other Coupling Constants of the Unified SuperStandard Theory* (Pingree Hill Publishing, Auburn, NH, 2019).

______, 2019b, *Coupling Constants of the Unified SuperStandard Theory SECOND EDITION* (Pingree Hill Publishing, Auburn, NH, 2019).

______, 2019c, *New Hybrid Quantum Big_Bang–Megaverse_Driven Universe with a Finite Big Bang and an Increasing Hubble Constant* (Pingree Hill Publishing, Auburn, NH, 2019).
______, 2019d, *The Universe, The Electron and The Vacuum* (Pingree Hill Publishing, Auburn, NH, 2019).

______, 2019e, *Quantum Big Bang – Quantum Vacuum Universes (Particles)* (Pingree Hill Publishing, Auburn, NH, 2019).

______, 2019f, *The Exact QED Calculation of the Fine Structure Constant Implies ALL 4D Universes have the Same Physics/Life Prospects* (Pingree Hill Publishing, Auburn, NH, 2019).

______, 2019g, *Unified SuperStandard Theory and the SuperUniverse Model: The Foundation of Science* (Pingree Hill Publishing, Auburn, NH, 2019).

______, 2020a, *Quaternion Unified SuperStandard Theory (The QUeST) and Megaverse Octonion SuperStandard Theory (MOST)* (Pingree Hill Publishing, Auburn, NH, 2020).

______, 2020b, *United Universes Quaternion Universe - Octonion Megaverse* (Pingree Hill Publishing, Auburn, NH, 2020).

______, 2020c, *Unified SuperStandard Theories for Quaternion Universes & The Octonion Megaverse* (Pingree Hill Publishing, Auburn, NH, 2020).

______, 2020d, *The Essence of Eternity: Quaternion & Octonion SuperStandard Theories* (Pingree Hill Publishing, Auburn, NH, 2020).

______, 2020e, *The Essence of Eternity II* (Pingree Hill Publishing, Auburn, NH, 2020).

______, 2020f, *A Very Conscious Universe* (Pingree Hill Publishing, Auburn, NH, 2020).

______, 2020g, *Hypercomplex Universe* (Pingree Hill Publishing, Auburn, NH, 2020).

______, 2020h, *Beneath the Quaternion Universe* (Pingree Hill Publishing, Auburn, NH, 2020).

______, 2020i, *Why is the Universe Real? From Quaternion & Octonion to Real Coordinates* (Pingree Hill Publishing, Auburn, NH, 2020).

Eddington, A. S., 1952, *The Mathematical Theory of Relativity* (Cambridge University Press, Cambridge, U.K., 1952).

Fant, Karl M., 2005, *Logically Determined Design: Clockless System Design With NULL Convention Logic* (John Wiley and Sons, Hoboken, NJ, 2005).

Feinberg, G. and Shapiro, R., 1980, *Life Beyond Earth: The Intelligent Earthlings Guide to Life in the Universe* (William Morrow and Company, New York, 1980).

Gelfand, I. M., Fomin, S. V., Silverman, R. A. (tr), 2000, *Calculus of Variations* (Dover Publications, Mineola, NY, 2000).

Giaquinta, M., Modica, G., Souchek, J., 1998, *Cartesian Coordinates in the Calculus of Variations* Volumes I and II (Springer-Verlag, New York, 1998).

Giaquinta, M., Hildebrandt, S., 1996, *Calculus of Variations* Volumes I and II (Springer-Verlag, New York, 1996).

Gradshteyn, I. S. and Ryzhik, I. M., 1965, *Table of Integrals, Series, and Products* (Academic Press, New York, 1965).

Heitler, W., 1954, *The Quantum Theory of Radiation* (Claendon Press, Oxford, UK, 1954).

Huang, Kerson, 1992, *Quarks, Leptons & Gauge Fields 2nd Edition* (World Scientific Publishing Company, Singapore, 1992).

Jost, J., Li-Jost, X., 1998, *Calculus of Variations* (Cambridge University Press, New York, 1998).

Kaku, Michio, 1993, *Quantum Field Theory*, (Oxford University Press, New York, 1993).

Kirk, G. S. and Raven, J. E., 1962, *The Presocratic Philosophers* (Cambridge University Press, New York, 1962).

Landau, L. D. and Lifshitz, E. M., 1987, *Fluid Mechanics 2nd Edition*, (Pergamon Press, Elmsford, NY, 1987).

Misner, C. W., Thorne, K. S., and Wheeler, J. A., 1973, *Gravitation* (W. H. Freeman, New York, 1973).

Rescher, N., 1967, *The Philosophy of Leibniz* (Prentice-Hall, Englewood Cliffs, NJ, 1967).

Rieffel, Eleanor and Polak, Wolfgang, 2014, *Quantum Computing* (MIT Press, Cambridge, MA, 2014).

Riesz, Frigyes and Sz.-Nagy, Béla, 1990, *Functional Analysis* (Dover Publications, New York, 1990).

Sagan, H., 1993, *Introduction to the Calculus of Variations* (Dover Publications, Mineola, NY, 1993).

Sakurai, J. J., 1964, *Invariance Principles and Elementary Particles* (Princeton University Press, Princeton, NJ, 1964).

Streater, R. F. and Wightman, A. S., 2000, *PCT, Spin, Statistics, and All That* (Princeton University Press, Princeton, NJ 2000).

Weinberg, S., 1972, *Gravitation and Cosmology* (John Wiley and Sons, New York, 1972).

Weinberg, S., 1995, *The Quantum Theory of Fields Volume I* (Cambridge University Press, New York, 1995).

Weinberg, S., 2000, *The Quantum Theory of Fields Volume III Supersymmetry* (Cambridge University Press, New York, 2000).

Weyl, H., 1950, *Space, Time, Matter* (Dover, New York, 1950).

Weyl, H., (Tr. S. Pollard et al), 1987, *The Continuum* (Dover Publications, New York, 1987).

INDEX

Anthropic Principle, 46
Arrow of Time, 73, 80, 81, 83, 117
Asynchronous Logic, 116, 128
axioms, 47
B.P. Abbott *et al*,, 23
Bailin, 59
baryonic force, 128
Big Bang, 53, 115, 117, 126, 127
Bjorken, J. D., 115
Black Hole, 115
BMOST, 113
Bogoliubov transformations, 70, 71
Bondi-Gold-Hoyle-Narlikar Steady State Cosmology, 22
bosonic string, 59, 60
BQUeST, 1, 3, 8, 13, 14, 31, 32, 35, 53, 99, 113
bright patches, 21
Charmonium, 31, 69, 127
Chomsky, 46, 48
cloaked propagator, 60
coherent state, 75, 76, 82
Cold Spot, 22
Complex General Relativity, 82, 116, 117
complex quaternion space, 10, 37
Complex Special Relativity, 72
Complexon Feynman Propagator, 62
condensation, 54
conservation law, 56
Cosmic Microwave Background, 21
Cosmological Constant, 117
Cosmos, iv, 119, 127
Coulomb gauge, 66, 71
coupling constants, 31, 69, 85, 86
CQMechanics, 68
Creation, 119
Dark Energy, 53, 116
Dark Matter, 116, 117
Dimensional Interaction, 54
dimensions, 54
Dirac, 73, 76, 77, 83
divergences, 60, 126
Durham University group, 22
eigenvalue function, 25, 27
ElectroWeak, 115, 116
energy-time uncertainty relation, 64
equal time commutation relations, 57, 66
Fermion group, 39, 40, 41, 85
fermion triangle divergence, 68
Feynman propagators, 54, 58, 60, 79, 92
fine structure constant, 127
Fine Structure Constant, 25
gauge invariance, 55, 56, 60
Gaussian, 54, 68
Generation group, 12, 40, 72, 82, 85, 92, 93
gravitational field fluctuations, 64
gravitational mass, 81, 82
graviton, 65
Great Attractor, 21
Gupta-Bleuler gauge, 56
Higgs Mechanism, 117, 126
Higgs microscope, 73
Higgs particles, 31, 69, 72, 73, 76, 78, 79, 80, 82, 83, 126
Hoyle and Narliker, 22
Hubble Constant, 22, 23
Hubble Parameter, 23, 24
imaginary coordinates, 53, 54, 59
indices, 4
inertial mass, 81, 82
inertial reference frames, 77, 78
interactions, 126
Interdimensional Interaction, 54

internal dimensions, 4, 100, 113
internal symmetries, 35
ISIS, 128
J. T. Nielsen *et al*,, 23
Jacobian, 55, 61
JBW, 27, 28
Johnson-Baker-Willey, 25, 27
K. Aylor *et al*, 22
Killing vector, 31, 69
Klein-Gordon field, 58
lagrangian, 55, 56
Laniakea Galaxy Supercluster, 21
Layer group, 12, 72, 82, 85, 92, 93
LHC, iv, 116
M. Soares-Santos *et al*, 23
mass scale, 54
massless QED, 25
Megaverse, iv, 92, 119, 120
MoND, 92
MOST, iv, 50, 108, 111, 119
Negative Energy Scalar Particle States, 76
negative energy states, 83
nonterminal, 46
Octonion, iv, 119
Parity Violation, 116
Pauli Exclusion Principle, 73
perturbation theory, 54
Planck mass, 54
plancktor field, 65
principle value propagators, 79, 80
Production Rules, 46
Pseudoquantum Field Theory, 31, 69
PseudoQuantum Field Theory, 68
QED, 63
Quantum, 48, 119, 126, 128
quantum computers, 126
Quantum Dimensions, 54, 59, 115
Quantum Entanglement, 117
quantum fluctuations, 54
Quantum Gravity, 54, 63, 64, 115
quark, 126
qube, 47
QUeST, iv, 10, 37, 119, 129
radiation gauge, 56, 57
Ranga-Ram Chary, 21
real dimensions, 8, 104
Reality group, 82
renormalization, 63
retarded propagators, 79, 80
Ricci tensor, 90, 91
Riemann-Christoffel Curvature Tensor, 48
Robertson-Walker metric, 126
scalar field, 57, 63
seed fermion, viii, 3, 4, 5
Shapley Attractor, 21
Space-Time Dependent Particle Masses, 81
Special Relativity, 115
species, 40, 41, 108, 111
spin, 126
spinor connection, 92
Standard Model, 60, 63, 115, 126
SU(3), 116
SuperStandard Model, iv, 118, 119, 120
Superstring, 59, 60, 115
SuperString theory, 113
SuperSymmetry, 118
symbols, 46, 48
symmetry breaking, 35
T Invariance, 78
T. Shanks et al, 22
terminal, 46
Theory of Everything, 72, 82, 115, 117
Thermodynamics, 116
tiles, x, 35, 38, 39, 41
time intervals, 64
translational invariance, 56
two-tier, 60, 61, 63, 64, 65, 90
Two-Tier coordinates, 53
two-tier formalism, 55
Two-Tier propagators, 59
Two-Tier PseudoQuantum Field Theory, 71
Two-Tier Quantum Field Theory, 53

U(4), 116, 117
U(8), 117
Unified SuperStandard Model, iv, 119, 120
universe eigenvalue function, 27
universe particle, 68
urfermion, x, 99, 100, 101, 113
UST, 129
UTMOST, 104, 105, 107, 108, 109, 110, 111, 112
vacuum fluctuations, 63, 65
vacuum polarization, 25, 26, 28
Vitamorphic Principle, 46
Weyl, H., 121
Wheeler-Feynman Electrodynamics, 92

About the Author

Stephen Blaha is a well-known Physicist and Man of Letters with interests in Science, Society and civilization, the Arts, and Technology. He had an Alfred P. Sloan Foundation scholarship in college. He received his Ph.D. in Physics from Rockefeller University. He has served on the faculties of several major universities. He was also a Member of the Technical Staff at Bell Laboratories, a manager at the Boston Globe Newspaper, a Director at Wang Laboratories, and President of Blaha Software Inc. and of Janus Associates Inc. (NH).

Among other achievements he was a co-discoverer of the "r potential" for heavy quark binding developing the first (and still the only demonstrable) non-Aeolian gauge theory with an "r" potential; first suggested the existence of topological structures in superfluid He-3; first proposed Yang-Mills theories would appear in condensed matter phenomena with non-scalar order parameters; first developed a grammar-based formalism for quantum computers and applied it to elementary particle theories; first developed a new form of quantum field theory without divergences (thus solving a major 60 year old problem that enabled a unified theory of the Standard Model and Quantum Gravity without divergences to be developed); first developed a formulation of complex General Relativity based on analytic continuation from real space-time; first developed a generalized non-homogeneous Robertson-Walker metric that enabled a quantum theory of the Big Bang to be developed without singularities at $t = 0$; first generalized Cauchy's theorem and Gauss' theorem to complex, curved multi-dimensional spaces; received Honorable Mention in the Gravity Research Foundation Essay Competition in 1978; first developed a physically acceptable theory of faster-than-light particles; first derived a composition of extremums method in the Calculus of Variations; first quantitatively suggested that inflationary periods in the history of the universe were not needed; first proved Gödel's Theorem implies Nature must be quantum; provided a new alternative to the Higgs Mechanism, and Higgs particles, to generate masses; first showed how to resolve logical paradoxes including Gödel's Undecidability Theorem by developing Operator Logic and Quantum Operator Logic; first developed a quantitative harmonic oscillator-like model of the life cycle, and interactions, of civilizations; first showed how equations describing superorganisms also apply to civilizations. A recent book shows his theory applies successfully to the past 14 years of history and to *new* archaeological data on Andean and Mayan civilizations as well as Early Anatolian and Egyptian civilizations.

He first developed an axiomatic derivation of the form of The Standard Model from geometry – space-time properties – The Unified SuperStandard Model. It unifies all the known forces of Nature. It also has a Dark Matter sector that includes a Dark ElectroWeak sector with Dark doublets and Dark gauge interactions. It uses quantum coordinates to remove infinities that crop up in most interacting quantum field theories and additionally to remove the infinities that appear in the Big Bang and generate inflationary growth of the universe. It shows

gravity has a MOND-like form without sacrificing Newton's Laws. It relates the interactions of the MOND-like sector of gravity with the r-potential of Quark Confinement. The axioms of the theory lead to the question of their origin. We suggest in the preceding edition of this book it can be attributed to an entity with God-like properties. We explore these properties in "God Theory" and show they predict that the Cosmos exists forever although individual universes (or incarnations of our universe) "come and go." Several other important results emerge from God Theory such a functionally triune God. The Unified SuperStandard Theory has many other important parts described in the Current Edition of *The Unified SuperStandard Theory* and expanded in subsequent volumes.

Blaha has had a major impact on a succession of elementary particle theories: his Ph.D. thesis (1970), and papers, showed that quantum field theory calculations to all orders in ladder approximations could not give scaling deep inelastic electron-nucleon scattering. He later showed the eigenvalue equation for the fine structure constant α in Johnson-Baker-Willey QED had a zero at α = 1 not 1/137 by solving the Schwinger-Dyson equations to all orders in an approximation that agreed with exact results to 4th order in α thus ending interest in this theory. In 1979 at Prof. Ken Johnson's (MIT) suggestion he calculated the proton-neutron mass difference in the MIT bag model and found the result had the wrong sign reducing interest in the bag model. These results all appear in Physical Review papers. In the 2000's he repeatedly pointed out the shortcomings of SuperString theory and showed that The Standard Model's form could be derived from space-time geometry by an extension of Lorentz transformations to faster than light transformations. This deeper space-time basis greatly increases the possibility that it is part of THE fundamental theory. Recently, Blaha showed that the Weak interactions differed significantly from the Strong, electromagnetic and gravitation interactions in important respects while these interactions had similar features, and suggested that ElectroWeak theory, which is essentially a glued union of the Weak interactions and Electromagnetism, possibly modulo unknown Higgs particle features, be replaced by a unified theory of the other interactions combined with a stand-alone Weak interaction theory. Blaha also showed that, if Charmonium calculations are taken seriously, the Strong interaction coupling constant is only a factor of five larger than the electromagnetic coupling constant, and thus Strong interaction perturbation theory would make sense and yield physically meaningful results.

In graduate school (1965-71) he wrote substantial papers in elementary particles and group theory: The Inelastic E- P Structure Functions in a Gluon Model. Phys. Lett. B40:501-502,1972; Deep-Inelastic E-P Structure Functions In A Ladder Model With Spin 1/2 Nucleons, Phys.Rev. D3:510-523,1971; Continuum Contributions To The Pion Radius, Phys. Rev. 178:2167-2169,1969; Character Analysis of U(N) and SU(N), J. Math. Phys. 10, 2156 (1969); and The Calculation of the Irreducible Characters of the Symmetric Group in Terms of the Compound Characters, (Published as Blaha's Lemma in D. E. Knuth's book: *The Art of Computer Programming Vols. 1 – 4*).

In the early 1980's Blaha was also a pioneer in the development of UNIX for financial, scientific and Internet applications: benchmarked UNIX versions showing that block size was critical for UNIX performance, developing financial modeling software, starting database benchmarking comparison studies, developing Internet-like UNIX networking (1982) and developing a hybrid shell programming technique (1982) that was a precursor to the PERL programming language. He was also the manager of the AT&T ten-year future products development database. His work helped lead to commercial UNIX on computers such as Sun Micros, IBM AIX minis, and Apple computers.
In the 1980's he pioneered the development of PC Desktop Publishing on laser printers and was nominated for three "Awards for Technical Excellence" in 1987 by PC Magazine for PC software products that he designed and developed.
Recently he has developed a theory of Megaverses – actual universes of which our universe is one – with quantum particle-like properties based on the Wheeler-DeWitt equation of Quantum Gravity. He has developed a theory of a baryonic force, which had been conjectured many years ago, and estimated the strength of the force based on discrepancies in measurements of the gravitational constant G. This force, operative in D-dimensional space, can be used to escape from our universe in "uniships" which are the equivalent of the faster-than-light starships proposed in the author's earlier books. Thus travel to other universes, as well as to other stars is possible.
Blaha also considered the complexified Wheeler-DeWitt equation and showed that its limitation to real-valued coordinates and metrics generated a Cosmological Constant in the Einstein equations.
The author has also recently written a series of books on the serious problems of the United States and their solution as well as a book on the decline of Mankind that will follow from current social and genetic trends in Mankind.

In the past twenty years Dr. Blaha has written over 80 books on a wide range of topics. Some recent major works are: *From Asynchronous Logic to The Standard Model to Superflight to the Stars, All the Universe!, SuperCivilizations: Civilizations as Superorganisms, America's Future: an Islamic Surge, ISIS, al Qaeda, World Epidemics, Ukraine, Russia-China Pact, US Leadership Crisis, The Rises and Falls of Man – Destiny – 3000 AD: New Support for a Superorganism MACRO-THEORY of CIVILIZATIONS From CURRENT WORLD TRENDS and NEW Peruvian, Pre-Mayan, Mayan, Anatolian, and Early Egyptian Data, with a Projection to 3000 AD,* and *Mankind in Decline: Genetic Disasters, Human-Animal Hybrids, Overpopulation, Pollution, Global Warming, Food and Water Shortages, Desertification, Poverty, Rising Violence, Genocide, Epidemics, Wars, Leadership Failure.*

He has taught approximately 4,000 students in undergraduate, graduate, and postgraduate corporate education courses primarily in major universities, and large companies and government agencies.

Recently he developed a quantum theory, The Unified SuperStandard Theory (UST), which describes elementary particles in detail without the difficulties of conventional quantum field theory. He found that the internal symmetries of this theory could be exactly derived from a 32 dimension complex quaternion theory called QUeST. He further found that a 32 dimension complex octonion theory (MOST) describes the Megaverse. It can hold QUeST universes such as our own universe. It has an internal symmetry structure which is a superset of the QUeST internal symmetries.

CPSIA information can be obtained
at www.ICGtesting.com
Printed in the USA
JSHW020721270920
8268JS00002B/48

9 781735 679501